T0339892

SIMULATION OF POWER SYSTEM WITH RENEWABLES

SIMULATION OF POWER SYSTEM WITH RENEWABLES

LINASH KUNJUMUHAMMED

HVDC Engineer, Mitsubishi Electric Europe, UK

STEFANIE KUENZEL

Lecturer in Power Systems, Royal Holloway University of London, UK

BIKASH PAL

Professor of Power Systems, Imperial College London, UK

Academic Press is an imprint of Elsevier
125 London Wall, London EC2Y 5AS, United Kingdom
525 B Street, Suite 1650, San Diego, CA 92101, United States
50 Hampshire Street, 5th Floor, Cambridge, MA 02139, United States
The Boulevard, Langford Lane, Kidlington, Oxford OX5 1GB, United Kingdom

Copyright © 2020 Elsevier Inc. All rights reserved.

MATLAB® is a trademark of The MathWorks, Inc. and is used with permission. The MathWorks does not warrant the accuracy of the text or exercises in this book.

This book's use or discussion of MATLAB® software or related products does not constitute endorsement or sponsorship by The MathWorks of a particular pedagogical approach or particular use of the MATLAB® software.

No part of this publication may be reproduced or transmitted in any form or by any means, electronic or mechanical, including photocopying, recording, or any information storage and retrieval system, without permission in writing from the publisher. Details on how to seek permission, further information about the Publisher's permissions policies and our arrangements with organizations such as the Copyright Clearance Center and the Copyright Licensing Agency, can be found at our website: www.elsevier.com/permissions.

This book and the individual contributions contained in it are protected under copyright by the Publisher (other than as may be noted herein).

Notices

Knowledge and best practice in this field are constantly changing. As new research and experience broaden our understanding, changes in research methods, professional practices, or medical treatment may become necessary.

Practitioners and researchers must always rely on their own experience and knowledge in evaluating and using any information, methods, compounds, or experiments described herein. In using such information or methods they should be mindful of their own safety and the safety of others, including parties for whom they have a professional responsibility.

To the fullest extent of the law, neither the Publisher nor the authors, contributors, or editors, assume any liability for any injury and/or damage to persons or property as a matter of products liability, negligence or otherwise, or from any use or operation of any methods, products, instructions, or ideas contained in the material herein.

Library of Congress Cataloging-in-Publication Data
A catalog record for this book is available from the Library of Congress

British Library Cataloguing-in-Publication Data
A catalogue record for this book is available from the British Library

ISBN: 978-0-12-811187-1

For information on all Academic Press publications visit our website at https://www.elsevier.com/books-and-journals

Publisher: Joe Hayton
Acquisition Editor: Lisa Reading
Editorial Project Manager: Mariana Kühl Leme
Production Project Manager: Surya Narayanan Jayachandran
Cover Designer: Christian J. Bilbow

Typeset by TNQ Technologies

We thank our parents, spouses and children, for they make us who we are, support and motivate us.

Contents

About the authors

Linash Kunjumuhammed received the PhD degree from Imperial College London in 2012. He is currently working with Mitsubishi Electric Europe.

Stefanie Kuenzel received the PhD degree from Imperial College London, London, in 2014. She is currently Head of the Power Systems group and Lecturer with the Department of Electronic Engineering at Royal Holloway University of London. She is an editor for IEEE Transactions on Sustainable Energy and her research interests include renewable generation, smart metering and transmission, including HVDC.

Bikash Pal is Professor of Power Systems at the Department of Electrical and Electronic Engineering, Imperial College London, London. He is research active in dynamics, stability, estimation and control of power system dominated by renewable generations. At Imperial College London, he teaches various power system courses at the graduate and postgraduate level.

Preface

Electric power system being physically large, there is very little scope to build a prototype test case for experimentation for research and education purpose. Interconnected power network has evolved over the entire 20th century and continues to do so. Historically any expansion of the system is planned, designed, built and finally commissioned for operation. Upon commissioning, there is little scope to radically change the hardware; computer simulation based on model of the system takes a very important role in planning and design stage. Since 1960s, there have been many advancements in power system simulation to help power engineers understand the operating behaviour and response of the system in the field. In the 40s and 50s, power flow simulations were carried out through analogue network analyzer, synchronous machines connected to network through transient network analyser, and then through numerical algorithms in digital computer simulation and solutions, memory for data storage and speed of computation were the limitations; fast decoupled power flow was invented to overcome those limitation in the 70s. Around the same time, electromagnetic transient programme (EMTP) was developed for network transient and switching study. Various tools from applied numerical algebra were introduced for exploiting sparsity, ill-conditioned and stiff system. With current days computing speed and storage, simulation of very large power systems are routine, there are several vendors meeting the simulation need of the industry.

While the models of synchronous machine, transformer, networks, loads and other important components of the system are well-established and standardized, new technologies of generation from wind and solar are now important part to be integrated into the simulation framework that has evolved over the years. Most of the time power from wind and solar is represented as equivalent generation or negative load for high-level network planning study. Because the power output characteristics from solar and wind are very much influenced by their conversion and control – rather than treating them as pure negative demand – it is very important to understand the underlying mechanism of power production and how their connection to the network makes an impact on the operational behaviour of the interconnected networks.

Postgraduate students and researchers in power engineering require full models of the system they wish to simulatein order to gain insight of the system behaviour during normal and abnormal operation. Commercial programs frequently provide models in a block format. While this allows for the study of the behaviour of the components, however, black-box models restrict the scope for customization for altering the modelling structure easily. For this very reason, many students and researchers write and build power system models for simulations from first principle. Matlab Simulink has been a very widely used tool both in education and practise for such purpose. The benefits of such models to the researchers are huge. The researcher is aware of any limitations and assumptions made in the model and has full flexibility in the implementation and analysis of the system. The unavoidable situation with this approach is that it takes the researchers a long time to develop a full system representation, before they can pursue their main research question. The unique advantage is that they reinforce their understanding of power system modelling through this process.

Although many textbooks help researchers by providing details of modelling, this book offers a step-by-step approach to modelling for Matlab/Simulink implementation including all major components used in current power system operation. The contents of the book cover part of the syllabus in university postgraduate course in power engineering covering renewable energy topics. The book will be useful to set exercises or course work, as it includes example studies with expected results. The book also provides a valuable resource for those familiar with the conventional power system technology and wish to learn more about the modelling of some of the more recent technologies. To build a comprehensive model of a power system, a student does not have to refer to many sources to gather models for conventional generators, wind farms, solar plants, and control devices as various chapters of this book provide this with the required compatible models with worked examples. It is believed within a week a student will be able to build, assemble and run his/her power system simulation programme with the proper knowledge and understanding of the operating characteristic of electric power network of 21st century.

This book has resulted from the experience of the authors to conduct research utilizing Matlab as simulation platform. Results of worked-out examples and various parameters used are not fully reflective of any standard equipment available in the market. The interpretation and insights are completely of the authors and any component design based on the simulation programme described when used for product development may offer some unsatisfactory performance for which authors will not be liable. Authors have given a general flavour of power system simulation for university students.

CHAPTER ONE

Introduction

1.1 Power system – history of development (Kundur)

Electric power system has a fascinating history of evolution over 130 years. Such a long and successful legacy can be described through four distinct phases:

Viable (1880–1900): Following the invention of carbon filament lamps in 1879 by Thomas Alva Edison, Pearl Street station in New York was illuminated with a small DC system (¼ square mile area) in 1882. Many installations followed in the early and mid-1880s in the United States and continental Europe. It had two voltages (100 and 110 V) that reduced operating efficiency of the network as current flow over longer cables had losses in the conductor and drop in voltage at the load. Thus, it limited the size and spread of the network. Around this time, based on the principle of Faraday's law, a device was developed in Europe, which could transform voltage. With financial support from George Westinghouse, William Stanley worked on this device further to develop what we know as proper transformer. Inspired by this success, Westinghouse installed a commercial AC system in Pennsylvania in 1886 on a trial basis, and by 1887, more than 30 Westinghouse AC systems were in operation. Nicola Tesla invented three phase magnetic field and induction motors in 1886–87. George Westinghouse bought all his patents on AC generator, motor, transformer, transmission and distribution system and employed to further developing the products from these concepts. This breakthrough provided a full-scale industrial war between AC–DC. The battle between AC–DC continued for several years until AC system had the final triumph when 1893 World's Fair in Chicago was illuminated by Westinghouse with a dozen of 750 KW AC two-phase 60 Hz generators supplying 8000 arc lights and 130,000 incandescent lamps. The electrification arrangement in the fair turned the buildings into a 'city of light'. Later on, more AC systems were built with synchronous generator driven by steam and hydroturbine.

Scalable (1900–50): Usually the source of generation being coal or other fossil fuel, it was realized that to make electricity affordable, it must be generated in bulk. So larger size generators were designed and made. But the site

Simulation of Power System with Renewables
ISBN: 978-0-12-811187-1
https://doi.org/10.1016/B978-0-12-811187-1.00001-9
© 2020 Elsevier Inc.
All rights reserved.

of the fuel being away from community, the voltage drop and losses became a real challenge, paving the way for higher transmission voltage, larger capacity transformers, etc. By 1940, 500 MW single synchronous generator and 345 kV of transmission voltage became very common in North America. The schematic of a typical AC interconnected power system is shown in Fig. 1.1. Power is generated at large power station close to the

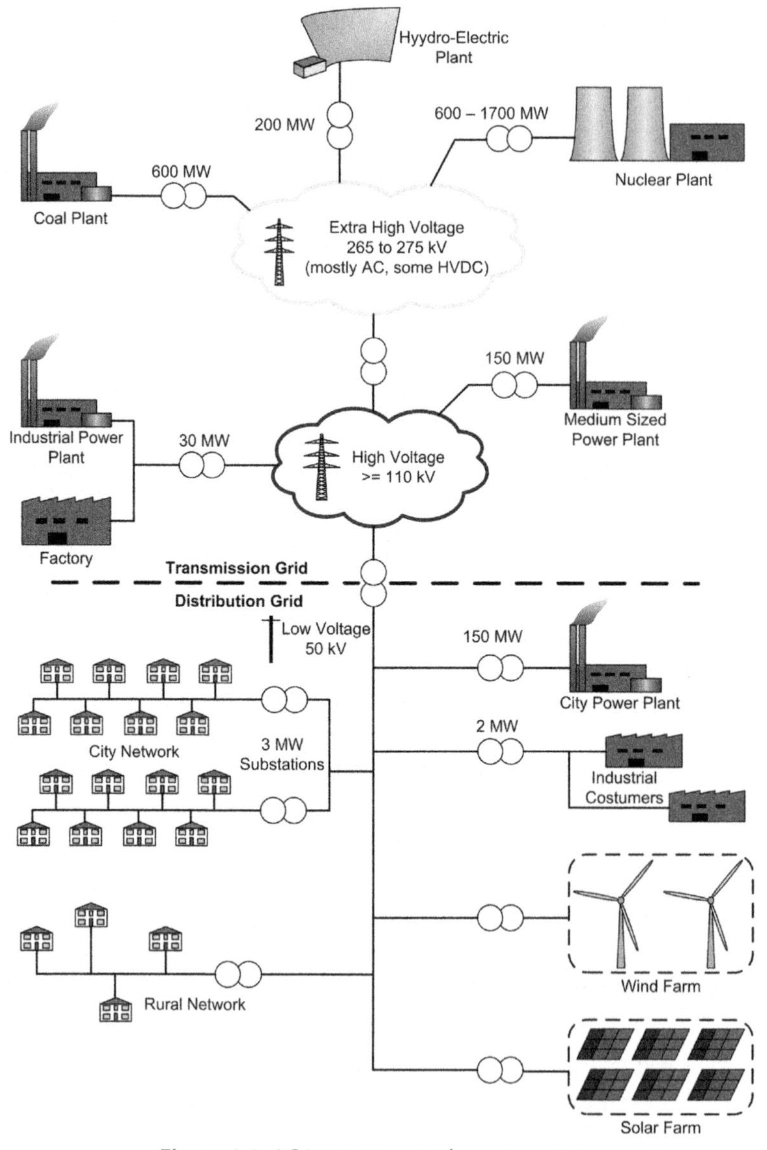

Figure 1.1 AC interconnected power system.

source of the fuel. Large synchronous generators are used to generate power at a voltage up to 25 kV. This is stepped up to transmission level voltage, which depends on the adopted standard value transmission voltage of a country, distance between the generation and demand, etc. Typical standard transmission voltages around the world are 220, 275, 345, 400, 500 and 765 kV. When the distance is far too long (about 1000 km or more), high-voltage DC transmission (HVDC) is used. HVDC is also used to connect to AC system of same or different operating frequency. At the consumers' end, the voltage is stepped down to subtransmission and distribution level at stages through step-down transformers. Usually, the consumer voltage is 400/415 V (line to line) or 110 V (mainly in the United States).

With such extra high voltage of transmission, the capacity and spread of the system grew enormously to connect almost every home and factory in the OECD countries. According to the survey of US National Academy of Engineering in 2000, North American electricity system was ranked the number one engineering achievement in 20th century as it is the largest man-made machine on the earth.

Reliable (1950—75): A 1000 MW power generator when connected to a network of hundreds of mile long 500 kV system through equivalent capacity transformers in power substations is subject to failure or shut down because of natural causes such as lightning stroke on the them, other operational failures such as certain quantities (e.g., temperature) exceeding design limits and so on. So their round-the-clock availability became very important. Interconnected power network of any reasonably sized country will have hundreds of such generators, transformers and thousands of km transmission circuits, and failure of any one of them should not lead to loss of supply to the customer. There have been several instances of system collapses (1965 power blackouts in the United States and many more after that throughout the world), so the reliability of the system became very important. Many countries have set up reliability council or group to come with operating practices and supply reliability standards, which are known as grid design and operation code. Although probability of failure of equipment can never be eliminated, how to maintain the demand despite outage of certain components was the direction of engineering innovations such as protection and relaying of the system. Improved design standard evolved.

Sustainable (1990-present): 66% of electricity generation in 2016 in OECD countries was from fossil fuel. When including non-OECD, this figure will go up further. For about 25 years, this percentage was even higher. Given that the gas reserve is predicted to last for another 50 years and the world will run out of coal in another 150 years with present rate of consumption, the source of energy has to move toward renewable form. The other important concern is the contribution of electricity generation into greenhouse gas emission. In 2015, it was estimated that in the United States, 29% of total greenhouse gas emission was contributed from its electricity generation. This figure for other large developing countries is even higher. So from the perspective of reduction in greenhouse gas emission, generation from low-carbon technology for the sustainability of the world economy and environment is important. There has been a massive growth of generation from renewables. In 2017 alone, 52 GW of wind and 95 GW of solar are added, taking the total installed capacity to 400 and 540 GW, respectively (WEC). Several countries have set targets of increasing percentage of generation from renewables. Already several countries in Europe such as Germany have several days at a stretch of electricity demand met from renewables only during low demand in summer. The power system is now facing challenges to support low-carbon generation with intermittency. Design, operation and control of power system are now undergoing significant innovation of unprecedented scale seen since 1940s.

Since the last phase of developments in power systems has been driven by a strong interest in renewable generation, power electronics has become indispensable part to understand, model and analyze. Power electronics is the enabling link between variable frequency wind generations and DC generation from PV. The widespread availability of communication, smart measurements and computation has further thrown interdisciplinary theme in the power system, which was previously unknown. The advances in power electronics, from the use of thyristors to IGBTs, have brought about many innovations. The classical current source converters had many limitations, which were overcome by the newer voltage source converter type. The increased interest in larger meshed DC grids has led to the search of a DC breaker technology. As mentioned, communication has played a major role in the development of power systems. Smart metres have been rolled out in many countries, to gain an improved observability and controllability of the demand side. Control of demand has become a part of the solution for future power systems, especially with an increasing proportion of electrified transportation.

Power system has evolved over past 130 years, and it will continue to evolve. Fig. 1.2 shows a time line.

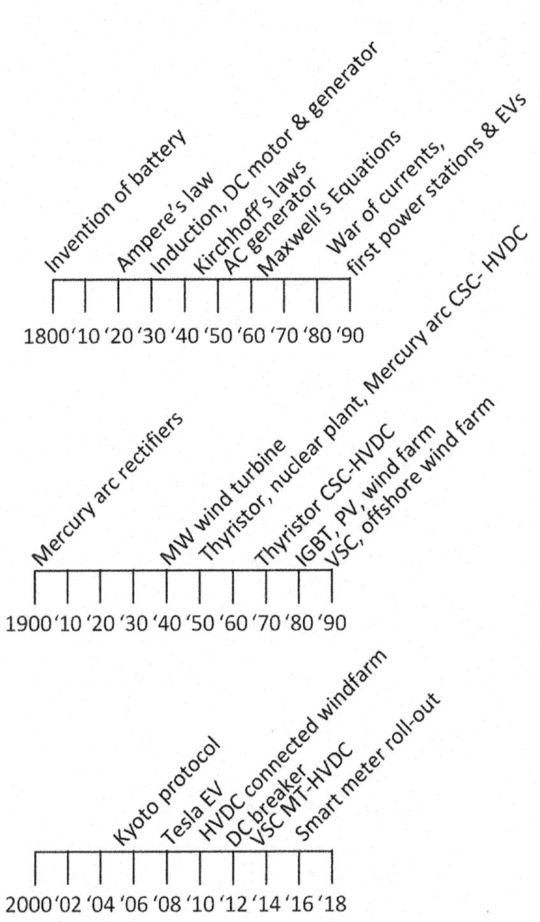

Figure 1.2 Power system development timeline.

1.2 Power system frequency

Power systems across the globe use nominal network frequencies of either 50 or 60Hz. The tolerance around this is within ±0.5%. It is very important to maintain the frequency as motoring loads and drives are designed to operate at a specified frequency; any notable variation will impact their performance. When the generated power is less than the demand and losses, the frequency will start to drop below the nominal frequency. When there is a surplus of generation, the frequency will start to rise above nominal. Fig. 1.3 shows typical frequency variation during a system frequency excursion event. The network frequency needs to be maintained within a tight bound around the nominal frequency. For this, the amount of

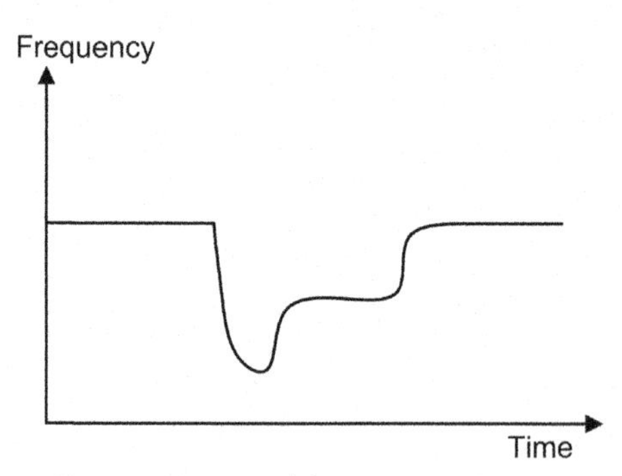

Figure 1.3 Typical frequency variation.

inertia in the system is very important. Inertia is a quantity that directly relates the amount of AC connected to rotating mass in the system. The rotating mass naturally opposes changes in frequency, by releasing kinetic energy during slow down and absorbing energy during speed up. Therefore, the inertia in the system determines how fast the system frequency can fall or rise (df/dt). If it takes 10 s to adjust generation levels to changes in demand, the df/dt will determine the lowest/worst value to frequency reaches. This point is called frequency nadir. In case of a generation surplus, there will be a frequency peak. Frequency is controlled through governor/gate control known as load frequency control (LFC), also known as primary frequency control; it is effective between 5 and 30 s, and it is automatic. But primary frequency control on its own will not be able to maintain the frequency to its preperturbed value. It is required to increase/decrease the fuel input set point to restore the frequency to its preperturbed value, is also automatic in many systems, takes about 30 s to few minutes and is known as secondary frequency response or LFC or automatic generation control; this is normally adjusting the volume of fuel for producing steam, for example, in coal-fired steam power plant. Further, slow tertiary control is normally provided manually through holding reserve in some selective units.

1.3 Phasors in AC systems

In AC Power system studies we deal with RMS values of voltage, current, not in instantaneous term as net power flow that we account are average value so we deal with phasor quantities, which can be described in different formats. In some situations, one format may be more convenient to work with than the other. Several formats of representing the same vector are frequently used in power systems.

One format is known as polar form. Using this format, the vector shown in the diagram below can be expressed by its magnitude (r) and phase angle (θ). The phase angle θ can be denoted in radians or degree. When a vector is expressed in exponential form ($re^{j\theta}$), the angle has to be in radians. As such, all the calculations using phase angles are performed in radians. Many of us are more familiar with angles in degree and have a better idea of what a 90 degree angle looks like than a $\pi/2$ angle. Hence, all the inputs and outputs are discussed and displayed in degree. The conversion in either direction is trivial. A circle has 360 degree or 2π radians. Converting 30 degree in radians, 30 degree*2π/360 degree $= \pi/6$.

Alternatively, the same vector can be described by complex numbers in Cartesian form with a real part (x) and imaginary part (y). We can conclude that the vector below can be described as $re^{j\theta}$, where θ is in radians or $x + jy$, where $x = r\cos\theta$, $y = r\sin\theta$. As we know from complex number theory, multiplication/division of two phasors is always convenient in polar form and addition and subtraction in Cartesian form – we will use this conveniently when modelling network, solving complex power flow equations, etc.

Quick check:

Convert $1 + 1j$ to polar form [sqrt(2) exp(j π/4)].

Give $2 + 0.5j$ in from of radius and angle in degrees [$r = 2.0616$, $\theta = 14.0362$ degree].

Converter 3 exp(j π 2/5) to Cartesian form [$0.9271 + 2.8532i$].

Give the vector of length 5 with angle 13 degree in Cartesian form [$4.8719 + 1.1248i$].

1.4 Per unit systems

We have learnt that power flow from generator to load takes through several steps of voltage transformation. Because current and voltages of either side of the transformers are different, impedance and admittance at one voltage level must be reflected to other voltage level through turns ratio adjustment. In a large network, it is really cumbersome, and one easily loses the handle. So we use normalization process, which is known as per unit conversions. Usually, base power is fixed at a value (typically 100 MVA), and operating voltage at each side is used as base voltage; this provides rated current value as base value on that side. The base impedance becomes ratio of base voltage/base current. The per unit impedance is the normalized impedance with respect to the base impedance. Because the voltage to current ratios of one side of the transformer is related to the other side by the square of the turn ratios, the normalized value of the impedance in either side of the transformer remains the same. This makes the calculation very convenient, so it is widely used in power system calculations. Following relations are useful to remember: $I_{\text{base}} = \frac{S_{\text{base}}}{V_{\text{base}}}$; $Z_{\text{base}} = \frac{V_{\text{base}}}{I_{\text{base}}}$; $Z_{\text{pu}} = \frac{Z_{\text{ohm}}}{Z_{\text{base}}}$; $Z_{\text{base}} = \frac{V_{\text{base}}^2}{S_{\text{base}}}$.pu conversion is widely used in power systems and not limited to electrical quantities. Further base values for conversion are introduced in Chapter 3, when modelling synchronous machines, with their rotating mass.

The line inductance and shunt capacitance are converted to impedance values first and then divided by base impedance to obtain respective per unit values, for series resistance and shunt conductance – their ohmic values are used for pu calculations.

The line reactance can be capacitive or inductive. Positive value represents inductive reactance, and negative value represents capacitive reactance. Generally symbol B is used for susceptance of the line and G is used for the conductance, both quantities are measured in Siemens and are entered as inputs. On per unit conversion it takes on the appropriate value. Bshunt and Gshunt are real and reactive shunt connected at the node. This means that Bshunt and Gshunt are susceptance and conductance values and like any others connected only at a bus rather than being part of the line. For both B and Bshunt, it holds true that if it is positive, if the line is capacitive, and negative if the line is inductive. In a line diagram, B is represented by two shunt elements of half the quantity.

1.5 Steady state in power system

AC power system operates with a fixed frequency (50 or 60 Hz), maintaining a balance between the supply and demand. The generation voltage is controlled in such a way that voltages at different points of the system are within an allowable range of designed value. As the demand changes, the generation follows it through its control, so there is always an operating equilibrium maintained through constancy of frequency. This is known as steady state, where all the equality constraints are met. Inequality constraints such as current and power flow in a feeder do not violate the limits; RMS values of voltages do not exceed the limits in either direction. The power flow equations are expressed as nonlinear static equations. Nonlinearity comes from the product of voltages and sine and cosine of phase angle between the voltages at different points. Usually unknown values of the voltages and angles are solved by solving the power flow equations iteratively because of the nonlinearity. Usually balanced operation is assumed so that single-phase representation can suffice. Normally at transmission level, the system is balanced, but it can be unbalanced at the distribution level, where three-phase power flow formulations should be used if the degree of unbalance exceeds some level. Whenever system is disturbed by way of loss of demand or generation, the system starts moving away from the equilibrium point at steady state. The temporal behaviour is governed by the dynamics of the system. We will discuss about that in detail, but it is important to understand that how the dynamics evolves will be dictated by initial operating points, so every dynamic situation will originate from a steady state.

From system theory point of view, in steady state, the derivatives of all the dynamic variables are zero. In other words, there is no change in any of the variables over time, also known as steady equilibrium point. A system in steady state can run for an infinite amount of time with unchanged variables, until a disturbance is introduced. A system can have a multitude of possible steady states. This concept is easily visualized in a scenario as shown in Fig. 1.4. Imagine a mountainous surface and a ball. The ball is in a steady-state operating point, where it can stay for an infinite amount of time, without any additional forces applied. It is obvious that the balls shown in red dashed lines are not in a steady-state operating point, as the ball will roll down the hill. Three of the green balls are in a good steady-state operating point, where they can stay until disturbed by an additional force. The forth green ball is in a critically stable position. The ball could rest at the top for an infinite amount of time; however, the smallest perturbation will suffice to push the ball out of the current operating condition. To summarize, the ball is in a steady-state operating point if the slope (derivative term) is zero.

1.6 Stability issues in power system

Stability is an inherent problem in interconnected power system operation. It is very complex in nature because of the temporal response of various components being different. Whenever disturbance appears in any part of the system, the balance between the generation and demand is created, which leads to moving away from the equilibrium. Depending

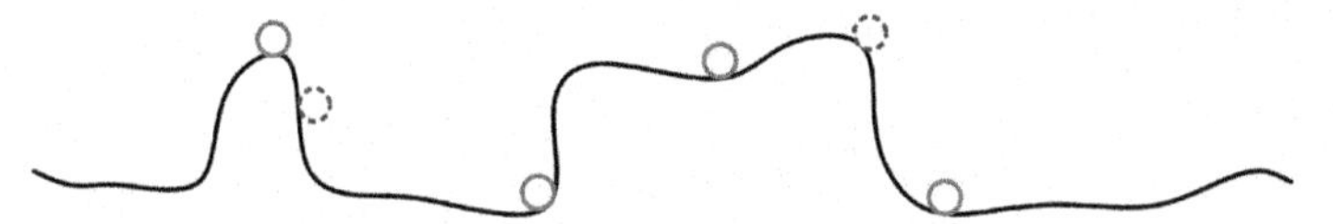

Figure 1.4 Stable and unstable states.

on the nature and location of the disturbance, the time scale of the dynamics evolved, as an example, a fault in the bus bar, will lead to large current and dropped voltage. If the bus bar is near the generator, the electrical power output from the generator will drop and the rotor will accelerate; this is known as transient stability problem and the time scale of observation is about 20 s. The concern is on the machine staying synchronous with the grid. On the other hand, if lightning is struck on a transmission circuit, very short duration (microseconds) high current (several kA) will result. Impact of lightning current is short duration, large voltage stress, which will damage the insulation. When gradually increased, load appears, the voltage drops, the system operates with reactive power from the generator and capacitor, not enough to hold the voltage at normal level; this is static voltage instability problem. With power converter interfacing renewable generation, other types of stability problem appear.

Nevertheless, the power system stability problem continues to be characterised in time scales and phenomena. Traditional lightning, switching surges in the network last between microseconds to hundreds of milliseconds. Added to this time scale is power converter switching dynamics. These can be classified as wave phenomena. The transients in generator stator, torsional vibration of generator shaft, swing of the rotor ranges between tens of milliseconds to few seconds – they are electromagnetic in nature at lower end and electrotechnical in nature at the higher end. Further slower dynamics represent governor and load frequency control in the range of tens of seconds to several minutes. Slow boiler dynamics and delayed voltage recovery of thermostatically control loads are in the time scale of hundreds of seconds to tens of minutes – mainly they are related thermodynamic process.

As in the past, the stability problem is being characterised based on the phenomena involving certain important parts and equipment of the system. Stability of the motion of rotor in synchronous machine is classified as rotor angle stability which is as ever electromechanical in nature. Reactive power imbalance driven time varying RMS voltage behaviour is classified as voltage stability – it is driven by both large and small disturbance. Their presence or consequence on the system can be short as well as long term. Elements involved in these cases are motors, generator excitation systems, slow transformer taps and mechanically switched capacitors. Balancing the system following sudden generation outage or demand connection is typically frequency stability problem. While frequency containment through generation spinning reserve and governor action is a primary frequency stability and control issue, bringing in more fuel inputs for increasing power production to meet the demand is a frequency stabilisation/restoration through secondary frequency control. Frequency and voltage stability these days are seen to be less and less decoupled of each other requiring new dynamic model and stability analysis and control tool. So, a clear boundary between pure frequency stability and pure voltage stability are fast disappearing. Integration of power converters as interface between networks and renewable generations is now a new piece of technology being increasingly deployed. There have been instances of dynamic excursion in RMS voltage and system frequency which can be purely electrical in nature and quite wide spread across the system involving several converters. This is a new form of stability/instability in interconnected power system operation – known as converter driven system stability. The dynamics have been slow as well as fast interacting – requiring further understanding through appropriate models beyond RMS phasor modelling framework.

Every equipment has its own control to safeguard the equipment from being stressed or damaged, like lightning strike on a transformer needs lightning arrestor and synchronous generator requires fast active automatic voltage regulators. The design of this control or protection requires study of the system when disturbed. Depending on the nature of the stability problem being investigated, suitable model is considered; normally, the study of power system under disturbance will require differential equations to describe the dynamic behaviour of the component under consideration in details and approximate model for the rest of the system. As an example, impact of fault on the generator bus bar terminal will require detailed dynamic model of generator and turbine and associated control to be modelled with behaviour of the rest of the system being modelled through algebraic approximation. The simulations of dynamics require initial operating condition, which are obtained with the help of power flow solution. While growth in renewable power system is good from sustainability and greenhouse gas reduction point of view, these new technologies of generation offer different set of operation and control challenges. The dynamic modelling of power converters and their switching control in the context of power network stability study are evolving. So there is a technical challenge to represent them in stability study of interconnected power system. In 2017, IEEE Power and Energy Society has commissioned a Task Force (TF) on: "Stability definitions and characterization of dynamic behavior in systems with high penetration of power electronic interfaced technologies". The TF is working on a report "Definition and Classification of Power System Stability Revisited". The report is expected to cover a detailed descriptions of stability phenomena, their associated time scales, various underlying causes and mechanisms and nature of manifestation with associated modelling and analysis tools to understand and address their impact on system operation.

1.7 Mathematical representation of power system

Power system model: As the time scale of dynamics is widely varying in nature, depending on the phenomena or time scale under study (say electromechanical stability), faster phenomena (such as power frequency transients) are represented through algebraic variables and very slower phenomena (as turbine steam and power flow) as fixed inputs. Accordingly, interconnected power systems are expressed through the following sets of equations:

$$\frac{d\boldsymbol{x}}{dt} = \boldsymbol{f}(\boldsymbol{x}, \boldsymbol{y}, \boldsymbol{u})$$

$$0 = \boldsymbol{g}(\boldsymbol{x}, \boldsymbol{y}, \boldsymbol{u})$$

$$\boldsymbol{z} = \boldsymbol{h}(\boldsymbol{x}, \boldsymbol{y}, \boldsymbol{u})$$

where, x is the vector of state variable vectors, y is the vector of algebraic variables, u is the vector of input variables and z is the vector of output variables. In the context of electromechanical oscillatory stability study, typically, generator speed, load angle, excitation system voltage and flux are time varying, which form part of state variables. The currents in the lines, voltage magnitude and angle are part of algebraic variables; generator input power and excitation system reference voltage are part of input variables. Depending on the control signal used, the output variable is chosen; it can be either state or algebraic variables. These vector equations are simulated in Matlab Simulink with known initial condition, normally obtained through power flow solution. For control design, normally, this system of equations is linearized around operating points, and algebraic variables ($\boldsymbol{y}$) are eliminated from all the equations to obtain linearized system in state space form or transfer function form as follow:

$$\Delta\dot{\boldsymbol{x}} = \boldsymbol{A}\boldsymbol{x} + \boldsymbol{B}\boldsymbol{u}$$

$$\boldsymbol{y} = \boldsymbol{C}\boldsymbol{x} + \boldsymbol{D}\boldsymbol{u}$$

$$\boldsymbol{G}(s) = \boldsymbol{C}(s\boldsymbol{I} - \boldsymbol{A})^{-1}\boldsymbol{B} + \boldsymbol{D}$$

where A, B, C and D are the state, input, output matrix and input to output coefficient matrices resulting from the linearized coefficient of the nonlinear differential, algebraic and output equation. The transfer matrix is obtained from state space following standard system theory approach. These linearized equations are very useful to analyze the stability property, obtain stability margin and design control for power system. In power system literature, this topic is also known as small signal stability.

1.8 Simulation in Matlab

Both Matlab and Simulink are products developed by Mathworks and are highly integrated. In this book, Matlab is the programming environment used to find all the initial parameters, to process and display data and to conduct further analysis. Simulink is used to build time domain models, containing the dynamics of the systems under study. This book assumes that users have experience in using Matlab. The user should be familiar with creating m-files and basic programming principles, such as functions and while, if and for loops. The use of Simulink is described in more detail in this book. If the user needs to freshen up on Matlab, good tutorial is provided in MathWork website.

In a power system, there are frequently several units, which are governed by the same equations, only with different parameters. One example for this is synchronous machines. In this case, vectors and matrices are a convenient way to model all parts of the system, which are alike in one single model. We set up the network in Simulink to fit with this vectorized approach (Fig. 1.5). Each of the models, which we describe throughout the book, takes system voltages as input and gives current injections as output. Every model, which contains several components, e.g., several generators, will take a voltage vector as input and give a current vector as output. The models are formulated in a format where vectors and matrices are used throughout, avoiding the need to duplicate the same equations multiple times for several equivalent components.

Dynamic power system models can be quite complex, and hence it is very convenient, being able to create systems made up of various subsystems. A model may, e.g., contain a subsystem for a synchronous generator, which in turn has a subsystem, which models the excitation system. If you look under tools, library browser and commonly used blocks, you can find a block called subsystem. You can drag and drop this into Simulink to create

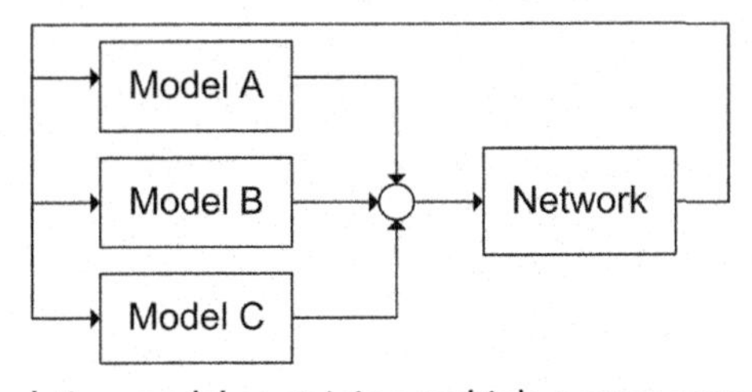

Figure 1.5 Simulation model containing multiple power system components.

a subsystem. You will see that by default this comes with an input and output, which are short circuited. You are able to alter this to the desired number of inputs and outputs by copy—paste, and further you can alter the relationship between the inputs and outputs within the subsystem. Create a subsystem in Simulink. Create three inputs and two outputs to the subsystem. You will see that these inputs and outputs inside the subsystem have numbers, determining the order in which they appear to the outside of the subsystem. By double clicking, you can change the port number and hence the order in which the inputs and outputs to a subsystem are listed. While this does not make a difference technically, this feature is very useful, to keep the model tidy and uncluttered as possible. All integrator blocks in Simulink need an initial condition, which is set by double clicking on the integrator block. In general, models are initialized to start in steady state. For this, all integrator blocks are set to the steady-state values. The steady-state values are calculated by setting all derivative terms to zero.

1.9 Assumptions

Various power system events have been discussed, together with the modelling requirements for their analysis. In the following chapters, this book will focus on power system dynamic stability studies for which following assumptions are reasonable:

- The network is symmetric: Self-impedance and mutual impedance of transmission lines are equally distributed in all the three lines. Voltages and currents in the line are balanced.
- Line parameters are constant or linear.
- Network is balanced. This allows single-phase representation of the three-phase system.
- Transformer: Represented using the π model; magnetizing reactance and phase shift introduced by star—delta transformer are usually neglected.
- Distribution systems consist of many types of loads, which are modelled as lumped loads — constant impedance, power or current types
- The system is under steady state at the beginning of the simulation. This helps to quantify the effect of a disturbance applied after a predefined time.

1.10 Summary

The chapter started with a brief history of the evolution of power system over past 130 years. It also talked about recent transition in terms of technology of generation, monitoring and control. Very essential but important aspect such as power system frequency, phasors and pu conversion are briefly introduced. The concept of steady state is explained. The notion of dynamics of power system over different time scales are introduced with various types of stability problem commonly encountered. It has also introduced the simulation platform (Matlab Simulink) and basic building blocks of representing each component in a vectorized framework. The chapter also listed standard assumptions about operation of the network and models of various components such as generators, transformers, transmission lines, etc.

Further reading

Kundur, P., 1994. Power System Stability and Control. McGraw Hill.

CHAPTER TWO

Transmission network modelling

This chapter sets out the simulation model of the backbone of the power system, the transmission network. An important part of this simulation program is the power flow (load flow) calculation to obtain the operating parameters of a transmission network such as magnitude and angle of bus voltages and active and reactive power flows across all branches. The chapter starts with the formulation of the admittance matrix, which relates bus voltages to bus current injections. The matrix is required for the computation of the power flow solution. Mathematical formulation of the power flow and an example solution using a three-bus system are illustrated next. The programming aspects of the power flow in computers and the code to obtain power flow solutions are elaborated at the end.

Once the power flow solution is obtained, the initial steady state operating points for other network components, such as synchronous machines, for the dynamic simulation can be easily calculated. The steady state defines the initial operating point at which the time derivative of all system states is zero. Time domain simulation usually starts from a steady-state operating condition. In this book, we will follow the steps presented in Fig. 2.1.

This chapter serves as guidance for those less familiar on the use of Matlab and Simulink, as it provides more details than later chapters. By the end of this chapter, the reader will be able to formulate an admittance matrix, find power flow solutions for small networks and be able to run a power flow simulation program and explain the results. Furthermore, the reader will know how to create and run a Simulink time domain simulation and validate the result. The network representation implemented in this chapter is the cornerstone to all the simulations in the later chapters.

2.1 Admittance matrix

Let us consider a simple network in Fig. 2.2, in which two transmission lines connect three buses. The lines are represented using the nominal π section model capturing the electrical properties of an overhead line such as resistance, inductance and capacitance. The shunt elements at Bus 2, shunt susceptance and shunt conductance, are represented using

Simulation of Power System with Renewables
ISBN: 978-0-12-811187-1
https://doi.org/10.1016/B978-0-12-811187-1.00002-0
© 2020 Elsevier Inc.
All rights reserved.

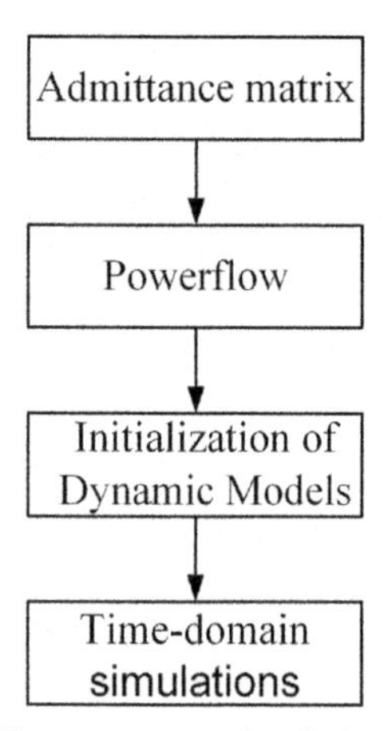

Figure 2.1 Power system simulation procedure.

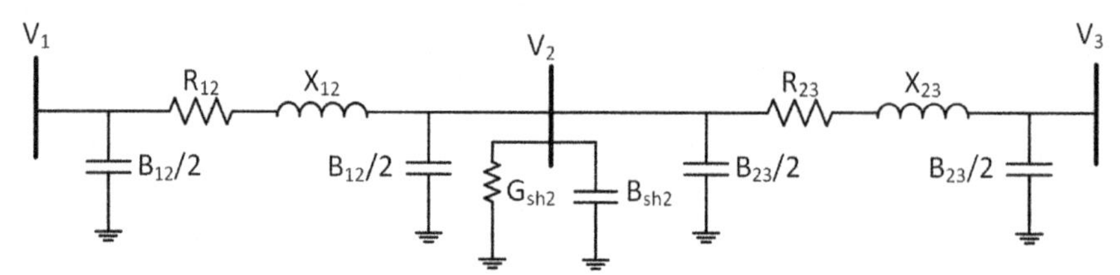

Figure 2.2 Representation of line and bus parameters.

B_{sh2} and G_{sh2}, respectively. The relationship between current injection at Bus 2 and the bus voltages is given by

$$I_2 = V_2\left(G_{sh2} + j\left(B_{sh2} + \frac{B_{12}}{2} + \frac{B_{23}}{2}\right)\right) + \frac{(V_2 - V_1)}{(R_{12} + jX_{12})} + \frac{(V_2 - V_3)}{(R_{23} + jX_{23})} \tag{2.1}$$

Self-admittance Y_{22} at Bus 2 is defined as

$$Y_{22} = \left(G_{sh2} + j\left(B_{sh2} + \frac{B_{12}}{2} + \frac{B_{23}}{2}\right)\right) + \frac{1}{(R_{12} + jX_{12})} + \frac{1}{(R_{23} + jX_{23})} \tag{2.2}$$

In addition, the mutual admittances Y_{21} and Y_{23} between buses 1 and 2 and buses 1 and 3, respectively, are defined as

$$Y_{21} = \frac{-1}{(R_{12} + jX_{12})} \tag{2.3}$$

$$Y_{23} = \frac{-1}{(R_{23} + jX_{23})} \tag{2.4}$$

Substituting (2.2–2.4) in (2.1), the equation for current injection at the Bus 2 becomes

$$I_2 = V_2 Y_{22} + V_1 Y_{21} + V_3 Y_{23} \tag{2.5}$$

Eq. (2.5) can be generalized for an n bus system as

$$\begin{bmatrix} \widetilde{I_1} \\ \widetilde{I_2} \\ \vdots \\ \widetilde{I_n} \end{bmatrix} = \begin{bmatrix} \widetilde{Y}_{11} & \widetilde{Y}_{12} & \cdots & \widetilde{Y}_{1n} \\ \widetilde{Y}_{21} & \widetilde{Y}_{22} & \cdots & \widetilde{Y}_{2n} \\ \vdots & \vdots & \ddots & \vdots \\ \widetilde{Y}_{n1} & \widetilde{Y}_{n2} & \cdots & \widetilde{Y}_{nn} \end{bmatrix} \begin{bmatrix} \widetilde{V_1} \\ \widetilde{V_2} \\ \vdots \\ \widetilde{V_2} \end{bmatrix} \tag{2.6}$$

$$\widetilde{\boldsymbol{I}} = \widetilde{\boldsymbol{Y}}\widetilde{\boldsymbol{V}} \tag{2.7}$$

where $\widetilde{\boldsymbol{Y}}$ is the admittance matrix, $\widetilde{\boldsymbol{V}}$ is bus voltage vector and $\widetilde{\boldsymbol{I}}$ is a current vector representing current injection at all buses. All the three quantities are complex values.

The $\widetilde{\boldsymbol{Y}}$ matrix can be formed by inspection from the line and bus parameters. The diagonal element $\widetilde{Y}_{kk}$ is the sum of admittances of all the elements connected at Bus k.

$$\widetilde{Y}_{kk} = \sum_{l=1}^{n} \widetilde{y}_{kl} + \sum_{l=1}^{n} \widetilde{y}_{Bkl} + \widetilde{y}_k \tag{2.8}$$

In the admittance matrix, the susceptance is added to the diagonal term of the bus it is connected to, leading to the term $\widetilde{y}_{Bkl}$. Buses may further have shunt elements installed, which are accounted for by the $\widetilde{y}_k$ term.

$$\widetilde{y}_{Bkl} = 0.5\, jB_{kl}, \widetilde{y}_k = G_{shk} + jB_{shk} \tag{2.9}$$

The off-diagonal element $\widetilde{Y}_{kl}$ is the negative value of the sum of admittances of all lines connecting Bus k and Bus l. The negative sign accounts for the negative sign in the receiving end voltage for every branch current calculation as seen in Eq. (2.1–2.5).

$$\widetilde{Y}_{kl} = -\widetilde{y}_{kl}, \widetilde{y}_{kl} = \frac{1}{R_{kl} + jX_{kl}} \tag{2.10}$$

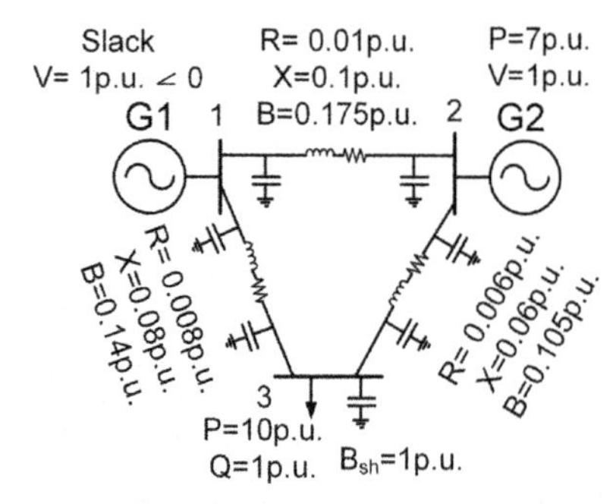

Figure 2.3 Two-machine, three-bus system.

2.2 Example

Let us understand the procedure of calculating an admittance matrix by inspection, using the three-bus system example shown in Fig. 2.3. Using the equations described above, find the admittance matrix for this system, checking your results as you go along with the provided solution.

The individual admittance values are

$$\widetilde{y}_{12}=\widetilde{y}_{21}=0.9901-j9.9010, \widetilde{y}_{13}=\widetilde{y}_{31}=1.2376-j12.3762,$$

$$\widetilde{y}_{23}=\widetilde{y}_{32}=1.6502-j16.5017$$

$$\widetilde{y}_{B12}=\widetilde{y}_{B21}=j0.0875,\ \widetilde{y}_{B13}=\widetilde{y}_{B31}=j0.07, \widetilde{y}_{B23}=\widetilde{y}_{B32}=j0.0525, \widetilde{y}_{3}=j1$$

The admittance matrix, containing the appropriate diagonal and off-diagonal terms, as shown below, leads to the following numerical end result:

$$Y=\begin{bmatrix} y_{12}+y_{13}+y_{B12}+y_{B13} & -y_{12} & -y_{13} \\ -y_{21} & y_{21}+y_{23}+y_{B21}+y_{B23} & -y_{23} \\ -y_{31} & -y_{32} & y_{31}+y_{32}+y_{B31}+y_{B32}+y_{3} \end{bmatrix}$$
$$=\begin{bmatrix} 2.2277-j22.1197 & -0.9901+j9.901 & -1.2376+j12.3762 \\ -0.9901+j9.901 & 2.6403-j26.2626 & -1.6502+j16.5017 \\ -1.2376+j12.3762 & -1.6502+j16.5017 & 2.8878-j27.7554 \end{bmatrix} \quad (2.11)$$

2.3 Power flow computation

The objective of the power flow program is to find the magnitude and angle of bus voltages given the active and reactive power load and active power generation at all buses. As losses are not known a priori, active power generation at one (sometimes more than one) of the buses, called slack bus or swing bus, is not specified. Instead, the magnitude and angle of voltage at the slack bus is specified usually (not necessarily) as 1 pu and zero degrees, respectively. There are four variables in each bus, voltage magnitude, voltage angle, active power and reactive power, and two of them are always specified. Accordingly, the buses are classified into the following three types:

Slack bus: This bus is the reference point for the rest of the AC network and is frequently chosen to be one of the generator buses. The bus voltage magnitude at this bus is commonly set to 1pu while the bus angle is set to 0 degree. As voltage magnitude and angle are provided as knowns, both the real power and reactive power at the bus are variables.

PV bus: All generator buses apart from the slack bus are commonly modelled as PV buses; however this is not a fixed rule. In the PV bus, the real power and voltage magnitude are known, while the reactive power and voltage angle are unknown.

PQ bus: A bus where active power and reactive power are fixed is called PQ bus. The magnitude and angle of voltage are unknowns. Buses that neither contain generation nor load naturally have zero entries for P and Q.

This section focuses on essential steps to obtain a power flow solution without going into details of the method. The three-bus, two-generator system shown in Fig. 2.3 is used as an example. With the admittance matrix at hand, which was calculated previously, we can find the load flow solution. Let us first go through a small derivation to understand the power flow equations.

The power injection at a bus is given by

$$P_k + jQ_k = \widetilde{V}_k \widetilde{I}_k^* \tag{2.12}$$

Multiplying out the matrix equation $\widetilde{\boldsymbol{I}} = \widetilde{\boldsymbol{Y}}\widetilde{\boldsymbol{V}}$ and taking the conjugate on both sides leads to

$$\widetilde{I}_k^* = \sum_{l=1}^{n} \widetilde{Y}_{kl}^* \widetilde{V}_l^* \tag{2.13}$$

Combining (2.12) and (2.13), where $\widetilde{Y} = G + jB$, gives

$$P_k + jQ_k = \widetilde{V}_k \sum_{l=1}^{n} (G_{kl} - jB_{kl}) \widetilde{V}_l^* \tag{2.14}$$

Converting the phasors $\widetilde{V}_k$ and $\widetilde{V}_l$ to polar form, we get

$$\widetilde{V}_k \widetilde{V}_l^* = V_k V_l (\cos \theta_{kl} + j \sin \theta_{kl}) \tag{2.15}$$

where $\theta_{kl} = \theta_k - \theta_l$.

Eq. (2.14) becomes

$$P_k + jQ_k = V_k \sum_{l=1}^{n} V_l (\cos \theta_{kl} + j \sin\theta_{kl})(G_{kl} - jB_{kl}) \tag{2.16}$$

The real and imaginary parts of (2.16) are

$$P_k = f_p(V, \theta) = V_k \sum_{l=1}^{n} V_l (G_{kl} \cos \theta_{kl} + B_{kl} \sin \theta_{kl}) \tag{2.17}$$

$$Q_k = f_q(V, \theta) = V_k \sum_{l=1}^{n} V_l (G_{kl} \sin \theta_{kl} - B_{kl} \cos \theta_{kl}) \tag{2.18}$$

Eqs. (2.17) and (2.18) can be written in the form given below:

$$\begin{bmatrix} P_k \\ Q_k \end{bmatrix} = \begin{bmatrix} f_p(V_1,\ V_2 \cdots V_n,\ \theta_1,\ \theta_2 \cdots \theta_n) \\ f_q(V_1,\ V_2 \cdots V_n,\ \theta_1,\ \theta_2, \cdots \theta_n) \end{bmatrix} \tag{2.19}$$

Let V and θ be the estimate of magnitude and angle of voltage at the beginning of an iteration, respectively, and ΔV and $\Delta\theta$ be the correction required to improve the solution. Taylor's theorem can be used to obtain the following result. Here, the details of derivation are skipped as the focus of the chapter is on implementing a power flow solution. Readers are advised to refer (Arrillaga, 2001) for more details.

$$\begin{bmatrix} P_k \\ Q_k \end{bmatrix} = \begin{bmatrix} f_p(V, \theta) \\ f_q(V, \theta) \end{bmatrix} = \underbrace{\begin{bmatrix} \dfrac{\partial f_p}{\partial \theta} & \dfrac{\partial f_p}{\partial V} \\ \dfrac{\partial f_q}{\partial \theta} & \dfrac{\partial f_q}{\partial V} \end{bmatrix}}_{j} \begin{bmatrix} \Delta\theta \\ \Delta V \end{bmatrix} \tag{2.20}$$

where J is called the Jacobian matrix. The Newton Raphson method using polar coordinates uses the inverse Jacobian matrix to update angles and voltage elements. Fig. 2.4 shows a flow chart for finding a power flow solution. The major step in the power flow calculation is the formulation of Jacobian matrix. Let us understand the formulation of this matrix. Once it is ready, we will return to the simple three-bus, two-generator system to

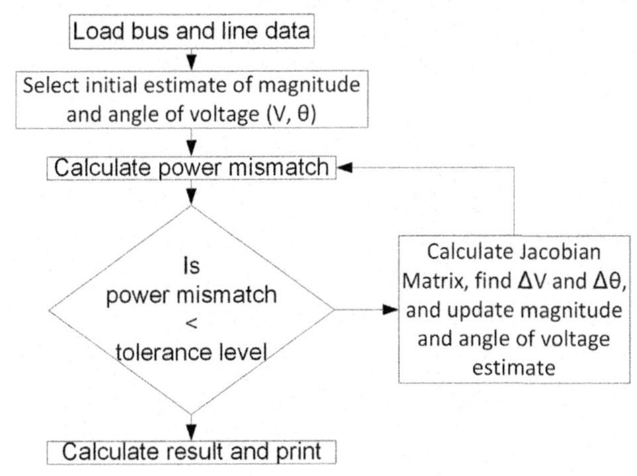

Figure 2.4 Program flow diagram of power flow solver.

practise the formulation of a Jacobian matrix and the procedure for the power flow solution.

2.4 Formulation of jacobian

The off-diagonal elements of J obtained using (2.17 and 2.18) are

$$\frac{\partial P_k}{\partial \theta_l} = V_k V_l (G_{kl} \sin \theta_{kl} - B_{kl} \cos \theta_{kl}) \tag{2.21}$$

$$V_l \frac{\partial P_k}{\partial V_l} = V_k V_l (G_{kl} \cos \theta_{kl} + B_{kl} \sin \theta_{kl}) \tag{2.22}$$

$$\frac{\partial Q_k}{\partial \theta_l} = - V_k V_l (G_{kl} \cos \theta_{kl} + B_{kl} \sin \theta_{kl}) \tag{2.23}$$

$$V_l \frac{\partial Q_k}{\partial V_l} = V_k V_l (G_{kl} \sin \theta_{kl} - B_{kl} \cos \theta_{kl}) \tag{2.24}$$

The diagonal elements of J are

$$\frac{\partial P_k}{\partial \theta_k} = - B_{kk} V_k^2 - Q_k \tag{2.25}$$

$$V_k \frac{\partial P_k}{\partial V_k} = P_k + G_{kk} V_K^2 \tag{2.26}$$

$$\frac{\partial Q_k}{\partial \theta_k} = P_k - G_{kk} V_k^2 \tag{2.27}$$

$$V_k \frac{\partial Q_k}{\partial V_k} = -B_{kk} V_k^2 + Q_k \tag{2.28}$$

An additional element, V_k or V_l, present in the left-hand side (LHS) of Eqs. (2.22), (2.24), (2.26) and (2.28) makes the right-hand side (RHS) of the equations similar, which makes calculation of J easier. The J matrix has to be calculated at every iteration, which demands an easy approach. A simple way of finding the elements of J matrix is to compute K using (2.29) for given estimate of magnitude and angle of voltage.

$$K = diag(V) Y^* diag\left(V^*\right) \tag{2.29}$$

Expanding (2.29), the off-diagonal and diagonal elements of K are obtained as (2.30) and (2.31), respectively.

$$K_{kl} = V_k V_l [(G_{kl} \cos \theta_{kl} + B_{kl} \sin \theta_{kl}) + j(G_{kl} \sin \theta_{kl} - B_{kl} \cos \theta_{kl})] \tag{2.30}$$

$$K_{kk} = V_k^2 (G_{kl} - jB_{kl}) \tag{2.31}$$

The real and imaginary parts of the off-diagonal elements of $\boldsymbol{K}$ are directly related to off-diagonal elements of J. A similar relation is visible with the diagonal elements of K and J.

For example, $\frac{\partial P_k}{\partial \theta_l} = imag(K_{kl})$ and $\frac{\partial P_k}{\partial \theta_k} = imag(K_{kk}) - Q_k$.

In this chapter, we will calculate the matrix K and pick elements of J through comparison.

2.5 Example of three-bus system

Let us apply the method in detail using the three-bus system example, which contains one PV bus, one PQ bus and one slack bus. By definition, PV buses have fixed active power generation and voltage magnitude. At these buses, terms corresponding to ΔV and ΔQ are absent. The Jacobian has two rows for each PQ bus and one row for each PV bus. No term appears for the slack bus, as both the voltage magnitude and angle are known.

As shown in Fig. 2.3, Bus 1 is the slack bus, where both the angle and voltage magnitude are known. This bus does not appear in the Jacobian. Bus 2 is a generator or PV bus, where the voltage angle is unkown. Only one row corresponding to ΔP appears for this bus. Bus 3 is a PQ bus, where neither the voltage nor angle is known, and two rows represent this bus in the Jacobian matrix.

The Jacobian for the study system is defined as

$$\begin{bmatrix} \Delta P_2 \\ \Delta P_3 \\ \Delta Q_3 \end{bmatrix} = \begin{bmatrix} \partial P_2/\partial\theta_2 & \partial P_2/\partial\theta_3 & \partial P_2/\partial V_3 \\ \partial P_3/\partial\theta_2 & \partial P_3/\partial\theta_3 & \partial P_3/\partial V_3 \\ \partial Q_3/\partial\theta_2 & \partial Q_3/\partial\theta_3 & \partial Q_3/\partial V_3 \end{bmatrix} \begin{bmatrix} \Delta\theta_2 \\ \Delta\theta_3 \\ \Delta V_3 \end{bmatrix} \tag{2.32}$$

The LHS is the real and reactive power mismatch vector for a given assumed voltage magnitude and angle. The vector at the RHS represents the required correction in the assumed voltage magnitude and angle.

Iteration 1:

Let the initial estimates for the three-bus system example be $V^0 = [1.0\ 1.0\ 1.0]$, $\theta^0 = [0.0\ 0.0\ 0.0]$, $P_2 = 7$, $P_3 = 10$ and $Q_3 = 1$, and the admittance Y has been calculated in (2.11). The essential calculations in the iteration are

$$\widetilde{I^0} = Y\widetilde{V^0} = \begin{bmatrix} 0.0 - 0.1575i \\ 0.0 - 0.1400i \\ 0.0 - 1.1225i \end{bmatrix}$$

$$P^0 + jQ^0 = \widetilde{V^0}\widetilde{I^{0*}} = \begin{bmatrix} 0.0 - 0.1575i \\ 0.0 - 0.1400i \\ 0.0 - 1.1225i \end{bmatrix}$$

Using (2.29),

$$K = \begin{bmatrix} 2.2277 + 22.1197i & -0.9901 - 9.9010i & -1.2376 - 12.3762i \\ -0.9901 - 9.9010i & 2.6403 + 26.2626i & -1.6502 - 16.5017i \\ -1.2376 - 12.3762i & -1.6502 - 16.5017i & 2.8878 + 27.7554i \end{bmatrix}$$

Comparing (2.21)–(2.28) with (2.30)–(2.31),

$$J = \begin{bmatrix} 26.4026 & -16.5017 & -1.6502 \\ -16.5017 & 28.8779 & 2.8878 \\ 1.6502 & -2.8878 & 26.6329 \end{bmatrix}$$

The readers are advised to compare the K and J matrix with Eqs. (2.21)–(2.31) to understand the formulation. At this point, $\Delta P_2 = P_2^0 - P_2 = -7$, $\Delta P_3 = P_3^0 - P_3 = -10$ and $\Delta Q_3 = Q_3^0 - Q_3 = 0.1225$. The corrections required in the initial estimate using (2.32) are $\Delta\theta_2 = 0.0758$, $\Delta\theta_3 = -0.2997$ and $\Delta V_3 = -0.0326$.

Iteration 2:

Improved estimates are $V^1 = [1.0\ 1.0\ 0.9674]$ and $\theta^1 = [0.0\ 0.0758\ -0.2997]$. These yield the following:

$$\widetilde{I^1} = Y\widetilde{V^1} = \begin{bmatrix} 2.8825 - 0.5295i \\ 6.8185 - 0.3632i \\ -9.3910 + 2.2273i \end{bmatrix}$$

$$P^1 + jQ^1 = \begin{bmatrix} 2.8825 + 0.5295i \\ 6.7715 + 0.8781i \\ -9.3161 + 0.6239i \end{bmatrix}$$

$$J = \begin{bmatrix} 25.3845 & -15.4370 & 4.5164 \\ -14.2661 & 25.3516 & -6.8363 \\ 7.3395 & -12.0187 & 27.4957 \end{bmatrix}$$

At end of this iteration, $\Delta P_2 = 0.2285$, $\Delta P_3 = -0.6839$ and $\Delta Q_3 = -1.6239$. The corrections required for the estimate are $\Delta\theta_2 = -0.0096$, $\Delta\theta_3 = -0.0540$ and $\Delta V_3 = -0.0801$.

Iteration 3:

New estimates are $V^2 = [1.0\ 1.0\ 0.8873]$ and $\theta^2 = [0.0\ 0.0662\ -0.3537]$.

$$\widetilde{I^2} = Y\widetilde{V^2} = \begin{bmatrix} 3.3584 - 1.6237i \\ 7.0794 - 1.8866i \\ -10.1021 + 4.7419i \end{bmatrix}$$

$$P^2 + jQ^2 = \begin{bmatrix} 3.3584 + 1.6237i \\ 6.9391 + 2.3507i \\ -9.8662 - 0.8423i \end{bmatrix}$$

$$J = \begin{bmatrix} 23.9119 & -13.9671 & 5.2202 \\ -12.7733 & 22.6948 & -8.5568 \\ 7.3060 & -12.1398 & 23.6784 \end{bmatrix}$$

At end of this iteration, $\Delta P_2 = 0.0609$, $\Delta P_3 = -0.1338$ and $\Delta Q_3 = -0.1577$. The corrections required in the estimate are $\Delta\theta_2 = -0.0014$, $\Delta\theta_3 = -0.0112$ and $\Delta V_3 = -0.0120$.

Iteration 4:

New estimates are $V^3 = [1.01.00.8753]$ and $\theta^3 = [0.0\ 0.0648\ -0.3649]$.

$$\widetilde{I^{*3}} = Y\widetilde{V^3} = \begin{bmatrix} 3.4521 - 1.7968i \\ 7.1507 - 2.1263i \\ -10.2612 + 5.1382i \end{bmatrix}$$

$$P^3 + jQ^3 = \begin{bmatrix} 3.4521 + 1.7968i \\ 6.9979 + 2.5850i \\ -9.9958 - 0.9965i \end{bmatrix}$$

$$J = \begin{bmatrix} 23.6776 & -13.7333 & 5.3743 \\ -12.5298 & 22.2635 & -8.8914 \\ 7.3307 & -12.2085 & 23.1571 \end{bmatrix}$$

At end of this iteration, $\Delta P_2 = 0.0021$, $\Delta P_3 = -0.0042$ and $\Delta Q_3 = -0.0035$. The corrections required in the estimate are $\Delta\theta_2 = -0.00003$, $\Delta\theta_3 = -0.0003$ and $\Delta V_3 = -0.0003$.

At the end of this iteration, the correction required in the estimate is very small. A couple of more iterations can produce even better results. Readers are advised to try further iterations to obtain a better estimate and also to gain an understanding that a large number of iterations are not often necessary. The reader may furthermore observe that a higher convergence rate is achieved during the first few iterations.

2.6 Power flow implementation

In the previous sections, we have learnt to calculate the power flow solution using simple steps for a three-bus system. However, for large networks, a power flow program is essential. In this section, two Matlab functions for calculating the admittance matrix and load flow are introduced. Both functions follow simple steps explained in the previous section and are not designed to guarantee the highest computational efficiency. They are tested for small networks and sufficient for building dynamic simulations discussed in this book.

There are many commercial and free power flow programs available. Any valid power flow solution will be sufficient to proceed following this book, and there is no need to change from a power flow solver the user

Table 2.1 Format of Bus matrix.

Bus No.	Voltage V	Angle δ	Generation P_G	Q_G	Load P_L	Q_L	G_{SHUNT}	B_{SHUNT}	Bus_id
1	1.02		5						2
2					5	1			3
3	1	0							1

Table 2.2 Format of Line matrix.

From bus	To bus	Resistance pu	Reactance pu	Line charging pu	Tap ratio
01	02	0.001	0.01	0.02	1

is already familiar with. If the user chooses to use another power flow program, extra care needs to be taken to name all outputs of the program according to the rest of the program and Simulink simulation.

As with any program, the power flow program used in this book expects input data to be presented in a specific format using bus matrix and line matrix. The network buses can be divided into generator (PV), load (PQ) and slack buses as previously discussed. They are identified using a parameter, bus_id = {1, 2, 3}. Load buses have the bus_id = 3 and generator buses have the bus_id = 2, while the slack bus has a bus_id = 1. Format of an example Bus matrix is given in Table 2.1. Compulsory elements are filled and others are optional or zero.

Similarly, the transmission lines connecting the bus bars also need to be defined in the program. The line data are presented using Line matrix as shown in Table 2.2. The lines are characterized by resistance, reactance, line charging and tap ratio.

The code for the matlab functions *form_Ymatrix*, which returns admittance matrix, and *power_flow*, which returns power flow solutions, are given below. The readers are advised to copy both the codes and save it as *form_Ymatrix.m* and *power_flow.m*, respectively, in a new folder and rename the folder Simulation. The program is tested using a four-machine test system example in the next section. The % symbol is used to provide nonexecuted comments in the Matlab code in Script 2.1 and 2.2.

2.7 Study case: four-machine system

The four-machine system in Fig. 2.5 is commonly studied in the power system community (Kundur, 1994). The system contains 4 generators G1−G4, 11 buses, 4 transformers, 8 transmission lines, 2 loads (indicated by arrows) and 2 shunt capacitances. Tables 2.3 and 2.4 list bus and line data of the system in their respective units.

```
function [Y] = form_Ymatrix(bs,ln)
% Form Y matrix from bus and line data

Y = zeros(size(bs,1),size(bs,1));  % Initialising Y bus to zero
sy = size(Y); % size of Y bus matrix

% This block of code adds line series admittance and shunt admittance to the diagonal
% elements of Y. First obtain index of diagonal elements corresponding to ln(:,1)
t = sub2ind(sy, ln(:,1), ln(:,1));
% Matrix Kt ensures correct admittance is added when more than one line connects two
% buses
[t1, t2, t3] = unique(t); Kt = zeros(size(t2,1),size(t3,1));
Kt(((1:size(Kt,2))-1)*size(Kt,1)+t3')=1;
% assigning values to diagonal elements
Y(t1) = Y(t1) + Kt*(1./(ln(:,3)+1i*ln(:,4))+1i*0.5*ln(:,5));

% This block does same as the previous block for the buses in ln(:,2)
t = sub2ind(sy, ln(:,2), ln(:,2));
[t1, t2, t3] = unique(t); Kt = zeros(size(t2,1),size(t3,1));
Kt(((1:size(Kt,2))-1)*size(Kt,1)+t3')=1;
Y(t1) = Y(t1) + Kt*(1./(ln(:,3)+1i*ln(:,4))+1i*0.5*ln(:,5));

% obtaining index of off-diagonal elements ln(:,1) to ln(:,2)
t = sub2ind(sy, ln(:,1), ln(:,2));
[t1, t2, t3] = unique(t); Kt = zeros(size(t2,1),size(t3,1));
Kt(((1:size(Kt,2))-1)*size(Kt,1)+t3')=1;
Y(t1) = Y(t1) - Kt*(1./(ln(:,3)+1i*ln(:,4)));

% obtaining index of off-diagonal elements ln(:,2) to ln(:,1)
t = sub2ind(sy, ln(:,2), ln(:,1));
[t1, t2, t3] = unique(t); Kt = zeros(size(t2,1),size(t3,1));
Kt(((1:size(Kt,2))-1)*size(Kt,1)+t3')=1;
Y(t1) = Y(t1) - Kt*(1./(ln(:,3)+1i*ln(:,4)));

% Adding shunt admittances at buses in the diagonal elements
Y = Y + diag(bs(:,8) + 1i*bs(:,9));

end
```

Script 2.1 Program to calculate Ymatrix.

2.8 Exercise

In Chapter 1, the benefits of handling power system calculations using a per unit system have been introduced. The readers are asked to convert the system parameters to the per unit system using the formulas discussed in Chapter 1 and compare it with the bus and line matrix provided in Script 2.3.

- Can you explain why the actual line data for the first four rows contain two entries for the inductance value, while the pu line data have only one entry?
- What is the source of the shunt capacitance in the line matrix?
- The capacitance connected at buses seven and nine are specified in MVar, not Farad. Why is this?

Now save script in Script 2.3 as *four_mac_data.m* in the *Simulation* folder. Run the m-file *four_mac_data*. The workspace on the Matlab main window should now show the bus and line matrices. Everything is in place to run the power flow for the four-machine test system shown in Fig. 2.5. Create an m-file named *four_mac_run.m* with the program in Script 2.4.

Run *four_mac_run.m*. The result, *bus_sol and line_flow* will be displayed in the command window. The slack bus for the simulation was set at Bus 3, where the real power and reactive power are the fourth and fifth element

```
function [bus_sln, flow] = power_flow(Y, bs, ln)

% The program solves load flow equations for a power system

bs(:,3) = bs(:,3)*pi/180; % converting angle from degrees to radians
V0 = bs(:,2); A0 = bs(:,3); nbs = size(bs,1);
bsl = find(bs(:,10)==1);   bpv = find(bs(:,10)==2);   ...
    bpq = find(bs(:,10)==3);

% Active and reactive power specified at PV and PQ buses

PQ = [bs(bpv,4)-bs(bpv,6);bs(bpq,4)-bs(bpq,6); bs(bpq,5)-bs(bpq,7)];

% Initial estimate of voltage angle (PV and PQ bus) and magnitude (PQ
bus)
vt0 = [bs(sort([bpv;bpq]),3);bs(bpq,2)];

itrn = 1;
while(true)
    % Updating voltage magnitude and angle
    T = [zeros(size(bpq,1),size(bpq,1)+size(bpv,1))
eye(size(bpq,1))];
    V0(bpq) = T*vt0;
    T = [eye(size(bpv,1)+size(bpq,1)) ...
        zeros(size(bpq,1)+size(bpv,1),size(bpq,1))];

    A0(sort([bpv; bpq])) = T*vt0;

% Calculate voltage, current and power based on the estimate

    v0 = V0.*exp(1i*A0);     i0 = Y*v0;   pq0 = v0.*conj(i0);

% Find difference in active and reactive power
    dpq = PQ-[real(pq0(sort([bpv;bpq])));imag(pq0(sort([bpq])))];

    K = diag(v0)*conj(Y)*diag(conj(v0));  % Calculating K matrix

    % Building Jacobian Matrix from K matrix
    Jp = [imag(K)-diag(imag(pq0)) ...
        (real(K)+diag(real(pq0)))./(ones(nbs,1)*V0')];

    Jq = [diag(real(pq0))-real(K) ...
        (imag(K)+diag(imag(pq0)))./(ones(nbs,1)*V0')];

    J = [Jp;Jq];
    J(sort([bsl; nbs+bsl;nbs+bpv]),:)=[];
    J(:,sort([bsl; nbs+bsl;nbs+bpv]))=[];

    dvt = J\dpq;     % Finding change in voltage magnitude and angle

    vt0 = vt0 + dvt;     % Updating voltage magnitude and angle

    itrn = itrn + 1;
    if max(abs(dvt))<1e-16 || itrn >50
        break
    end

end
```

Script 2.2 Program to calculate power flow.

```
% Building solved bus matrix
bus_sln = bs;
bus_sln(:,2:3) = [abs(v0) angle(v0)*180/pi];
bus_sln(sort([bpv;bsl]),4:5) = [real(pq0(sort([bpv;bsl]))) ...
    imag(pq0(sort([bpv;bsl])))];
bus_sln(sort([bpq]),6:7) = -[real(pq0(sort([bpq]))) ...
    imag(pq0(sort([bpq])))];

% calculating line flow
sln = size(ln,1);   flow = zeros(2*sln,4);
t = sub2ind(size(Y), ln(:,1), ln(:,2));
yft = Y(t);
flw = -v0(ln(:,1)).*conj(((v0(ln(:,1))-v0(ln(:,2))).*yft));
flow(1:sln, 1:4) = [ln(:,1) ln(:,2) real(flw) imag(flw)];

flw = -v0(ln(:,2)).*conj(((v0(ln(:,2))-v0(ln(:,1))).*yft));
flow(sln+1:2*sln, 1:4) = [ln(:,2) ln(:,1) real(flw) imag(flw)];
end
```

Script 2.2 Cont'd.

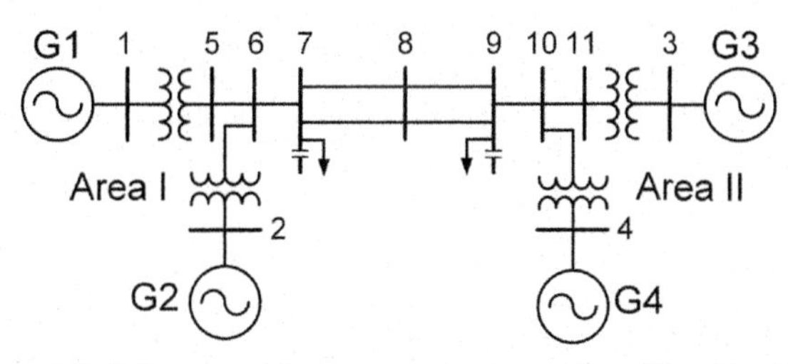

Figure 2.5 A four-machine, two-area test system (Kundur, 1994).

Table 2.3 Bus data of the four-machine, two-area system in actual values.

#	Bus V (kV)	Delta (degree)	Pgen (MW)	Qgen (MVar)	Pload (MW)	Qload (MVar)	Pshunt (MW)	Qshunt (MVar)	Bus type
1	20	0	700	0.0	0.00	0.0	0.0	0.0	PV
2	20	0	700	0.0	0.00	0.0	0.0	0.0	PV
3	20	0	0.0	0.0	0.00	0.0	0.0	0.0	Slack
4	20	0	700	0.0	0.00	0.0	0.0	0.0	PV
5	230	0	0.0	0.0	0.00	0.0	0.0	0.0	PQ
6	230	0	0.0	0.0	0.00	0.0	0.0	0.0	PQ
7	230	0	0.0	0.0	967	100	0.0	200	PQ
8	230	0	0.0	0.0	0.00	0.0	0.0	0.0	PQ
9	230	0	0.0	0.0	1767	100	0.0	350	PQ
10	230	0	0.0	0.0	0.00	0.0	0.0	0.0	PQ
11	230	0	0.0	0.0	0.00	0.0	0.0	0.0	PQ

Table 2.4 Line data of the four-machine, two-area system in actual values.

From bus	To bus	R (Ω)	L (mH)	C (μF)	Tap ratio	Tap phase (degree)
01	05	0.0000	23.4 HV (0.17719 LV)	0.000000	1.0	0.0
02	06	0.0000	23.4 HV (0.17719 LV)	0.000000	1.0	0.0
03	11	0.0000	23.4 HV (0.17719 LV)	0.000000	1.0	0.0
04	10	0.0000	23.4 HV (0.17719 LV)	0.000000	1.0	0.0
05	06	1.3225	35.1	0.219380	1.0	0.0
10	11	1.3225	35.1	0.219380	1.0	0.0
06	07	0.5290	14.0	0.087751	1.0	0.0
09	10	0.5290	14.0	0.087751	1.0	0.0
07	08	5.8190	154.4	0.965260	1.0	0.0
07	08	5.8190	154.4	0.965260	1.0	0.0
08	09	5.8190	154.4	0.965260	1.0	0.0
08	09	5.8190	154.4	0.965260	1.0	0.0

```
%four_mac_data.m % bus No., Voltage Mag., Voltage Angle, Pgen,
Qgen, Pload, Qload, Pshunt, % Qshunt, bus type
bus = [...
1 1.00    00.0   7.000     0.00 0.0000   0.000    0.00  0.00 2;
2 1.00    00.0   7.000     0.00 0.0000   0.000    0.00  0.00 2;
3 1.00    00.0   0.0000    0.00 0.0000   0.000    0.00  0.00 1;
4 1.00    00.0   7.0000    0.00 0.0000   0.000    0.00  0.00 2;
5 1.00    00.0   0.0000    0.00 0.0000   0.000    0.00  0.00 3;
6 1.00    00.0   0.0000    0.00 0.0000   0.000    0.00  0.00 3;
7 1.00    00.0   0.0000    0.00 9.6700   1.000    0.00  2.00 3;
8 1.00    00.0   0.0000    0.00 0.0000   0.000    0.00  0.00 3;
9 1.00    00.0   0.0000    0.00 17.670   1.000    0.00  3.50 3;
10 1.00   00.0   0.0000    0.00 0.0000   0.000    0.00  0.00 3;
11 1.00   00.0   0.0000    0.00 0.0000   0.000    0.00  0.00 3];

%from bus, To bus, Resistant, Inductance, Capacitance, tap-ratio,
tap-phase
line = [01 05 0.0000 0.0167 0.00000 1.0 0.0;
         02 06 0.0000 0.0167 0.00000 1.0 0.0;
         03 11 0.0000 0.0167 0.00000 1.0 0.0;
         04 10 0.0000 0.0167 0.00000 1.0 0.0;
         05 06 0.0025 0.0250 0.04375 1.0 0.0;
         10 11 0.0025 0.0250 0.04375 1.0 0.0;
         06 07 0.0010 0.0100 0.01750 1.0 0.0;
         09 10 0.0010 0.0100 0.01750 1.0 0.0;
         07 08 0.0110 0.1100 0.19250 1.0 0.0;
         07 08 0.0110 0.1100 0.19250 1.0 0.0;
         08 09 0.0110 0.1100 0.19250 1.0 0.0;
         08 09 0.0110 0.1100 0.19250 1.0 0.0];
```

Script 2.3 Script for bus and line matrices of the four-machine test system.

```
%four_mac_run.m
four_mac_data
[Y] = form_Ymatrix(bus,line);
[bus_sol, line_flow] = power_flow(Y,bus, line)
```

Script 2.4 Script for running the four-machine test system power flow.

in the row, respectively. If everything goes right, the slack bus active and reactive power generation will be 7.2343 pu and 1.6024 pu, respectively, for the data mentioned above. Alternatively the same information can be found in the workspace. The *line_flow* shows active and reactive power flows in the transmission lines.

2.9 Exercise

Note the power flow is different from bus A to bus B compared with bus B to bus A. For example, note the power flow between buses 1 to 5 and 5 to 1. Try to explain the reason. Then look at the power flow of buses 5 to 6 and 6 to 5 and explain.

Let us try some more exercises:

- Reduce active power demand (Pload) at Bus 7 to 867 MW. What is the new result? The slack bus should now produce 631.94 MW and 148.84 MVar. Why is it that with the load drawing 100 MW less power, the slack only produces 91.49 MW less?
- With no change in reactive power load, why is there a change in reactive power generation?
- Now make G1 slack bus, G3 generator bus and set G3 output equal 700 MW. What is the new result? The slack bus should now produce 618.65 MW and 132.77 MVar. Why is it that simply swapping the slack bus and nothing else gives a different result for the apparent power production?
- Try making further changes and see if you can explain the results. For further simulations, set the data back to the original values.

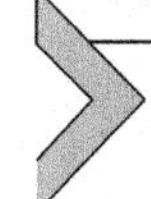

2.10 Including the network in the Simulink time domain simulation

The power flow solution is the starting point to almost any time domain simulation in power systems. With this first step successfully mastered, running a time domain simulation of the network is not very difficult. As mentioned in the introduction chapter, given that certain circumstances are satisfied, the network can be represented as an impedance matrix for time domain simulations. Chapter V discusses the modelling of loads in detail. In the meantime, we will make the simplifying assumption that the loads are of impedance type, and we will include all loads implicitly in the impedance matrix.

A few more lines need to be added at the end of *four_mac_run.m*, before we are ready to set up the Simulink model. Copy code in Script 2.5 and append it in the *four_mac_run.m*.

In the first instance, create a Simulink model, selecting *file* then *new* then *model* and save this as *Network_Timedomain*. In the new model, the library browser can be found under the view tab. The blocks you require can be dragged across to the new model from there. To build the model, follow the steps as indicated in Fig. 2.6, until you have the same system.

```
%Calculates the vector of apparent power S̃ injected by the
%generators for each system bus.

        PQ = bus_sol(:,4)+1i*bus_sol(:,5);

%Calculates the complex voltage value for all buses.

        vol = bus_sol(:,2).*exp(1i*bus_sol(:,3)*pi/180);

%Calculates the complex value of the current injected by the
%generators at each system bus

        Icalc = conj(PQ./vol);


%PL and QL are the real and reactive power drawn by loads at all
%system buses.

        PL = bus_sol(:,6);
        QL = bus_sol(:,7);

%V is the voltage magnitude at all system buses.

        V = bus_sol(:,2);

%The loads are assumed to be impedance type loads here and are
%included in the Ỹ matrix. For more information on modelling loads,
%see Chapter V. The Ỹ matrix is then inverted to find the impedance
%matrix.

        YPL = PL./V.^2;
        YQL = QL./V.^2;
        Y = Y + diag(YPL-j*YQL);
        Z = inv(Y);
```

Script 2.5 Program to calculate impedance matrix.

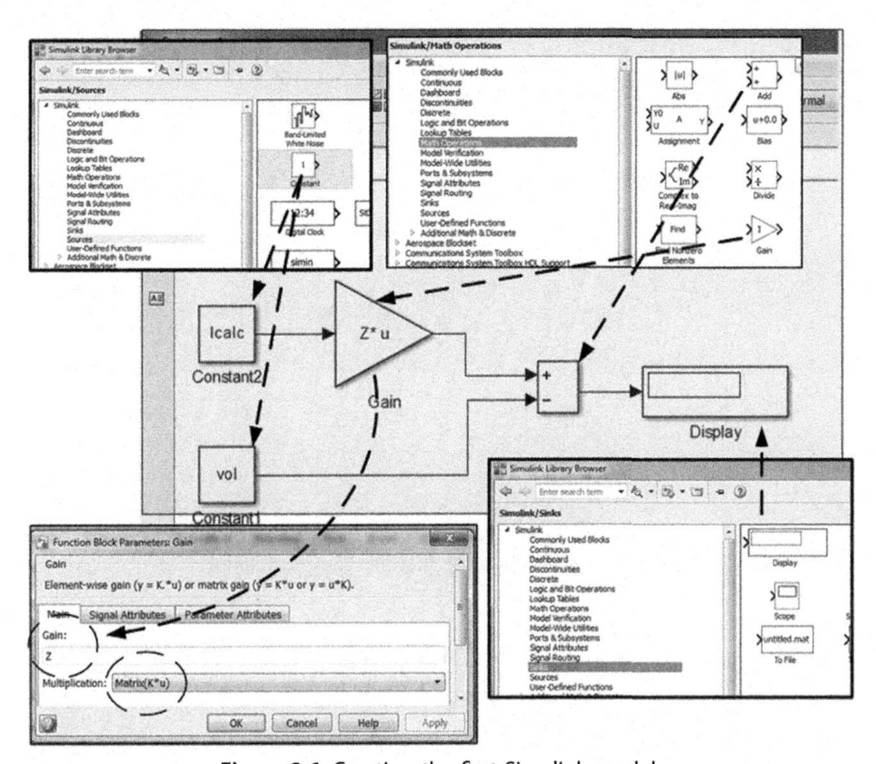

Figure 2.6 Creating the first Simulink model.

Once complete, click the run buttom, the black arrow on the top banner. If everything has been set up correctly, the output of the gain block should be vol; hence, vol-vol should be zero (or very, very small according to the precision of the calculation). The result from this first Simulink simulation can be seen in Fig. 2.7. We can see that for all bus voltages, the difference is 10^{-15} or smaller, which is zero for the purpose of the simulation. We will use this method for representing network in simulation models containing synchronous and asynchronous machines in the later chapters.

If this simulation throws up an error, check that the values from your power flow and initialization are still in the Matlab workspace. Open Block Parameters of the Gain block and ensure Multiplication option is set to Matrix (K*u). Similarly List of Signs of Add block is set to $\pm$. Ensure that the Simulink model is in the same root directory as the power flow and initialization. If in doubt, rerun the initialization and the Simulink model. If the display does not show all numbers close to zero, go back and validate your power flow results and check the initialization.

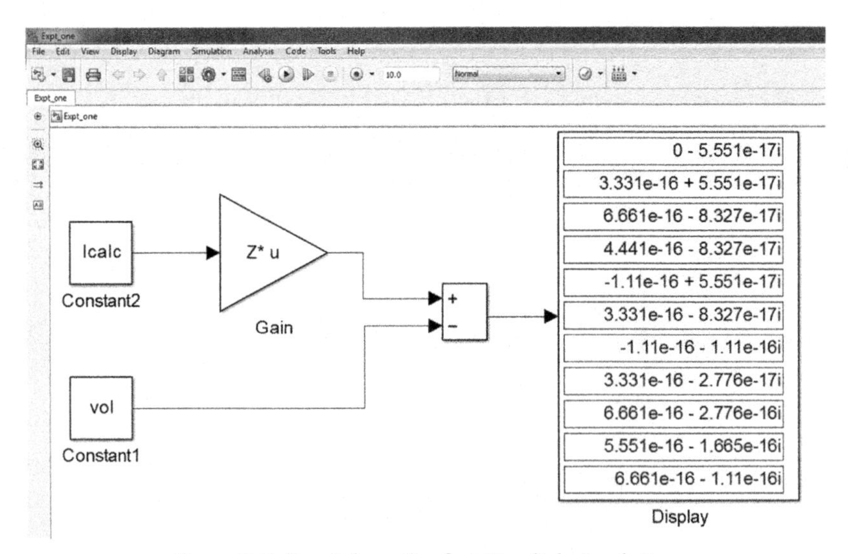

Figure 2.7 Result from the first Simulink simulation.

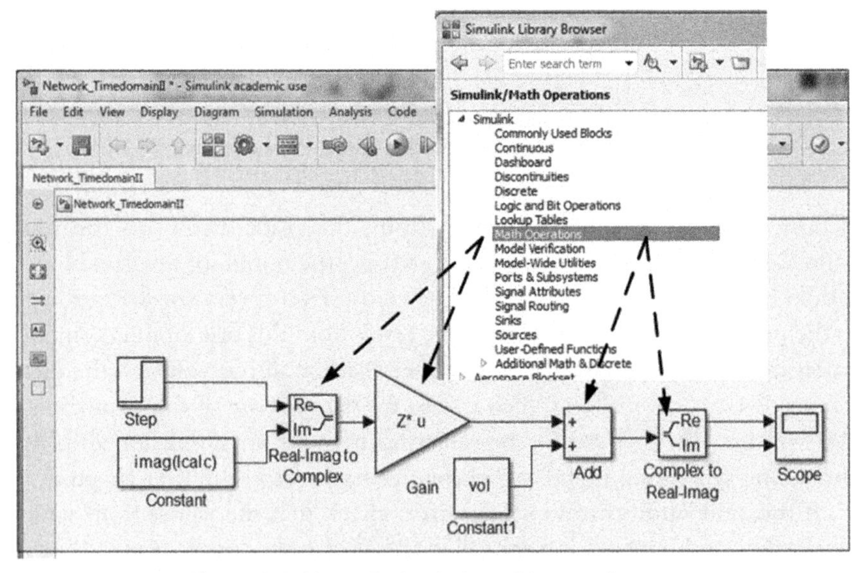

Figure 2.8 Network simulation with step change.

Is everything working so far? Time to introduce a step change. As before, follow the steps indicated in Fig. 2.8 to build a new model.

Under Sources in the library browser, there is a block called Step. Drag this into the model. Try setting the initial value to Icalc and click ok.

This should throw up an error message 'Invalid setting'. The reason for this is that certain Simulink blocks do not handle complex numbers. What to do now? The solution is to treat the real and imaginary parts separately. For this, four inbuilt functions are useful real(x), imag(x), abs(x) and angle(x). These provide the real part, imaginary part, magnitude and angle (in radians) of the complex number x. Let us say, we only want to introduce a step in the real part of Icalc, not in the imaginary part. The constant Icalc turns into imag(Icalc). Simple calculations like this can be typed directly into the Simulink model. The initial value for step is real(Icalc). Setting the final value of Icalc to real(Icalc)*2 leads to a 200% increase in the real part of Icalc at the time of the step. By default, this is set to 1 s. Now the real and imaginary parts need to be recombined to a complex number. You could either multiply the imaginary part by j and then add them or use another block from the library. In the math operations section, there is a block called 'real-imag to complex' to recombine the signals where no multiplication with the imaginary unit j is required. Another useful block from the library is the scope, found under sinks. Add this to your model. The scope, just like the step, does not take complex numbers as input. The real and imaginary parts or magnitude and angle can, however, be plotted separately. Use the Complex to Real-Imag block to split the signal. Double click on the scope and select the parameter symbol in the top ribbon (to the right of the printer symbol). In the general category, set the number of axis to two, and in the history selection, deselect the option 'limit data points to last:'. It is possible to stretch a box to make it larger. Simulink will show the name of the variable, e.g., in a constant block, as long as the box is large enough.

Run the model. Then double click the scope. This opens a window with the output from the scope, which will display plot as shown in Fig. 2.9.

2.11 Conclusions

This chapter has covered how to form the admittance matrix for a given network. The example used is a three-bus system with a PQ, PV and slack bus. The power flow computation has been discussed. For this purpose, the formation of the Jacobian matrix has been described, and the three-bus system was used as the example. The step-by-step iterative procedure of gaining increasingly accurate power flow solutions has been shown by using the numerical example of the three bus case. With the background behind the formation of the admittance and Jacobian matrix and the iterative power flow solution covered, a simple power flow program

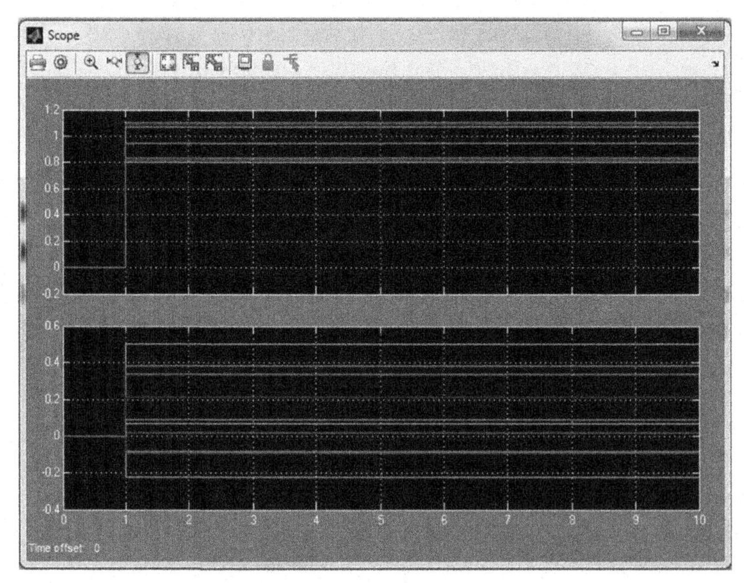

Figure 2.9 Scope showing step change in voltages.

has been introduced. In the example, the conversion of the four machine data from actual to pu data is covered, as the power engineer may be given data either in pu or in actual values. Hence, the conversion process is an important first step in the power flow calculation. Having attained a power flow solution of sufficiently high precision, the network can be modelled for time domain analysis in Simulink. Detailed guidance on this has been provided with snapshots of the simulation setup, as this is the first chapter using Simulink. Later chapters will assume a certain degree of familiarity of the user in building Simulink models. It was shown how to validate and troubleshoot the Simulink result and how to introduce a step change. You are now ready to go to the next chapter to learn how to integrate a detailed model of a conventional generator with the Simulink model of the network.

References

Arrillaga, J., Watson, N.R., 2001. Computer Modelling of Electrical Power Systems, Second Edition. John Wiley & Sons, Ltd.

Kundur, P., 1994. Power System Stability and Control. McGraw-Hill, Inc.

CHAPTER THREE

Synchronous machine modelling

3.1 Synchronous machine introduction

With ever-evolving history over the past 130 years, synchronous machine as technology (Neidhofer, 1992) has driven the development and progress of electrification in the planet through interconnected AC power network. The power networks in major countries each have thousands of synchronous generators producing power either at 50 or 60 Hz to keep the lights on. Both from the design and operational perspective, it is one of the major challenges to an engineer to model and analyze the machine. The dissipation of heat because of losses in the conductors and stresses on the insulation because of voltage have always challenged the engineer for going for large capacity individual machine with higher generation voltage. Accordingly, various currents and voltages are limited as defined by the generator capability curve. With continuous progress in cooling and insulation technologies, the capacity and constructional features of synchronous generator have continued to change but not the principle of synchronous generation. The current development trend is more on high-temperature superconductor technology for winding and air cooling for heat dissipation.

This chapter covers the simulation model for synchronous machines. First, the general operation of synchronous machines is described, detailing the differences between cylindrical rotor and salient pole machines. Next, a simple study system is introduced, the single machine infinite bus (SMIB) system. In the second section, the d-q and D-Q reference frames are introduced, which are essential for synchronous machine modelling. Dynamic equations for the machine modelling are provided. The model considers only the operation of the generator within its capability curve. Various limits such as rotor field heating current limit, stator heating current limit, armature core end iron heating limit, stability limit, etc., need to be modelled additionally. The initialization procedure to determine a steady-state operating point of the synchronous machine model is described, where the initialization is described both for system base and machine base. The Simulink implementation is described step by step, including the initial conditions. The Simulink section explicitly explains

Simulation of Power System with Renewables
ISBN: 978-0-12-811187-1
https://doi.org/10.1016/B978-0-12-811187-1.00003-2
© 2020 Elsevier Inc.
All rights reserved.

how to model the generator in machine base and system base. The reader is instructed into the art of debugging Simulink models in a methodical manner, leading to a working model. By the end of the chapter, the reader will be ready to run a time domain simulation with a synchronous machine in machine and system base and understand how the results compare. Various dynamic synchronous machine models are introduced, to raise awareness among the reader about the different modelling assumptions made. The chapter concludes with step-by-step guidance of building a dynamic simulation model for a two-area test system consisting of four synchronous machines.

3.2 Synchronous machine operation

In this subsection, we discuss some of the fundamental concepts related to synchronous machines and how they work. This understanding will help the reader to be able to follow the later sections of this chapter.

Fig. 3.1 shows the schematic diagram of a three-phase synchronous machine. The machine consists of a stator with armature windings and a rotor with field windings; the rotor and stator are separated by a small air gap just enough to have clearance. The direction of rotor rotation is indicated in the figure; the electrical speed of rotation ω is directly linked (synchronized) with the network frequency, which is the main characteristic of a

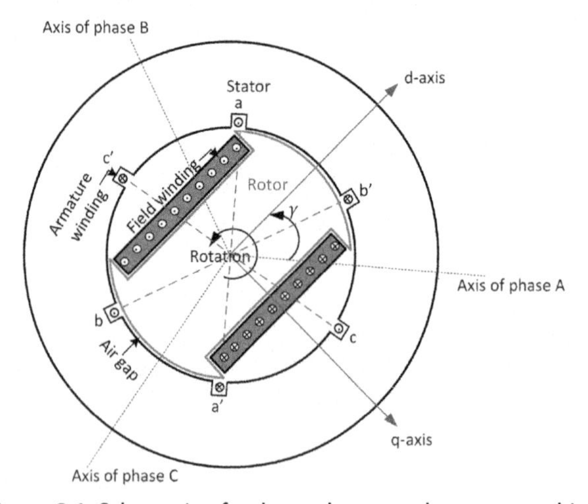

Figure 3.1 Schematic of a three-phase synchronous machine.

synchronous machine. In a machine with one pole pair, the mechanical rotational speed ω_m is the same as the electrical speed of rotation. In a machine with more pole pairs (n), the rotor needs to turn slower to achieve the same variation in the magnetic field, such that $\omega = \omega_m$ n. The field current in the field winding of the rotor is a DC current supplied by a field exciter. The armature windings are laid out, such that the three phases are 120 electrical degrees apart, where 360 electrical degrees is the angular distance between identical magnetic conditions. This definition is important in machines with multiple pole pairs. For the general understanding and development of equations, we can use a single-pole pair machine, without loss of generality. As the rotor rotates at an electrical speed ω, it can be naturally understood that the induced voltages in the armature are time varying with ω, with a phase difference of 120° between each phase.

Let us use the well-known right-hand rule to work out the meaning of the different axes in Fig. 3.1. The stator has three armature windings for the three phases, the first winding is between a and a', another winding is between b and b' and the third winding is between c and c', as indicated in Fig. 3.1. Using the right-hand rule, can one determine the reason for the direction of phase A, B and C axes, shown in Fig. 3.1? In the same way, we can investigate how the orientation of the field winding on the rotor is related to the direction of the axis named d-axis in Fig. 3.1.

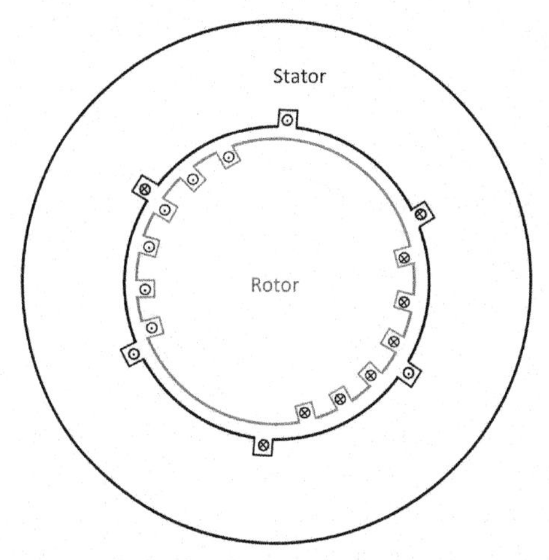

Figure 3.2 Cylindrical rotor synchronous machine.

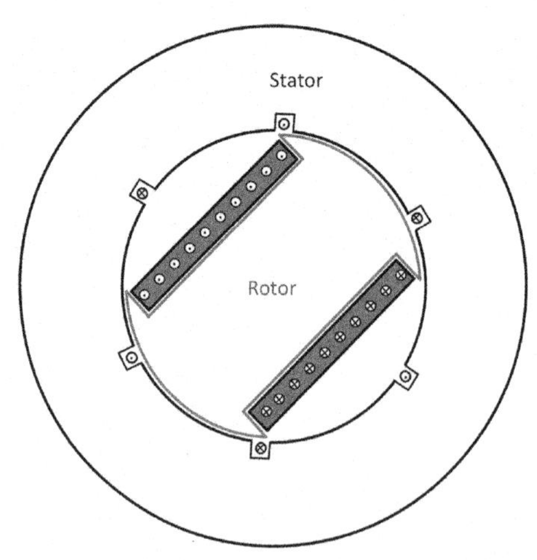

Figure 3.3 Salient pole synchronous machine.

There are two general types of rotors, cylindrical rotors and salient pole rotors. The difference between the two types can be clearly seen in Figs. 3.2 and 3.3.

It can be seen that for both the cylindrical and salient pole machine, there exist two axis of symmetry with respect to the rotor shape and air gap. These two axes of symmetry are the d-axis and q-axis, shown in Figure 3.1. As the rotor with the rotor circuits is rotating relative to the stator circuits, the time-varying flux path leads to self- and mutual inductances of the armature windings as well as mutual inductances between armature and rotor circuits, which are time varying, according to the rotor position (Clarke, 1957). Time-varying inductances will complicate the machine model, and resulting equations are computationally expensive to solve. The dq-transformation discussed in the next section is helpful to simplify the machine model.

3.3 Reference frame

In this section, we will see how the symmetry of the dq-axis can be used to analyze a machine, where the rotor circuits are rotating at a constant speed relative to the stator circuits. The dq-transformation avoids the need to work with time-varying inductances, which makes it a powerful tool for the analysis of synchronous machines. This section explains the mathematical transformation between the three-phase voltages of phase A, B and C and the dq0 representation. This section also covers the dq-reference frame of synchronous machine and the DQ-reference of the system, which is required for a network with several generators, which will operate in different dq-frames. The transition from one reference frame to the other is described as well.

Direct-quadrature-zero (dq0) transformation is frequently used in the analysis of three-phase circuits. As shown in Fig. 3.1, the direct component is aligned along the field axis (d-axis), which is perpendicular to the plane of the field windings. The quadrature component is aligned with the quadrature (q) axis, which lags the field axis by 90°. The d-axis leading the q-axis is IEEE convention. As the field axis is part of the rotor, it rotates at the speed ω in the direction indicated in the figure. The d-axis and q-axis rotate together with the rotor. In effect, the dq0 transformation changes the reference frame from a stationary frame to a rotating one, such that rotating phasors appear stationary from the perspective of this rotating reference frame. In any three-phase balanced system, the zero components are zero, such that we are left with a direct and quadrature component; hence, we will call the reference frame a dq-reference frame.

The readers should be aware at this point that there are multiple conventions in use. In simulations, care has to be taken to use the same convention consistently. This book uses the generator convention and power invariant park transform. The generator convention power injected into a network is positive, while power drawn from a network is negative. The orientation of the direct and quadrature axis has been extensively discussed in the literature. An IEEE Committee report (IEEE committee report, 1969) on phasor diagrams in 1969 concluded that it is preferred for the direct axis to lead the quadrature axis by 90°. 'Using this principle, the quadrature axis can be taken as the real axis. The direct axis then becomes the imaginary axis...'. In this book we follow this convention, such that $V_t = V_q + jV_d$. Fig. 3.4 shows the direct axis, quadrature axis, V_t, V_q and V_d.

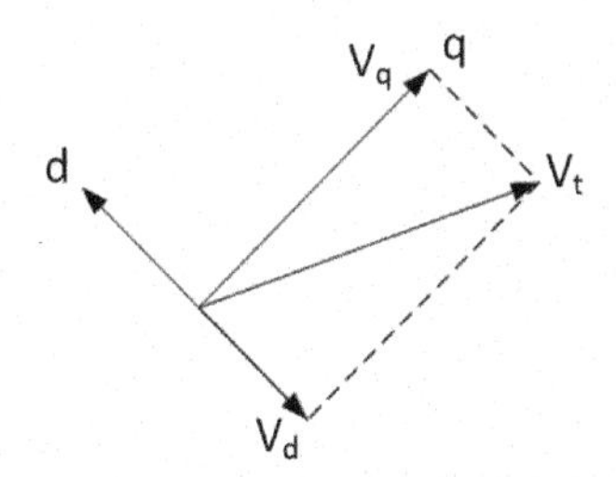

Figure 3.4 Phasor diagram of dq-frame.

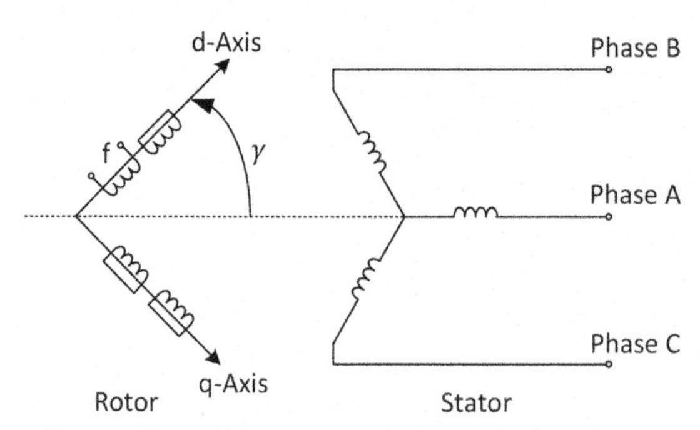

Figure 3.5 Showing position of dq-axis in relation with three-phase stator voltages.

The angle between the axis of phase A and the d-axis is denoted here as γ, as can be seen in Figs. 3.1 and 3.5. As the rotor and the dq-axis are turning at a speed ω, γ will be time varying at $\gamma = \gamma_0 + \omega t$. The field coil and one equivalent damper coil are on the d-axis, while there are two equivalent damper coils on the q-axis.

As the dq-reference frame is aligned with the field of the synchronous machine, it is easy to imagine how there may be several different dq-reference frames in a system, which has several synchronous machines. When simulating such a system, we need a common reference frame. This reference frame is the network reference frame, named DQ-reference frame. The DQ-reference frame in relation to the dq-reference frame is shown in Fig. 3.6.

When the study system contains an infinite bus, the slack bus will be the infinite bus by nature. If the infinite bus angle is zero degrees, the Q-axis of the DQ-frame is aligned with the infinite bus voltage; if the infinite bus has an angle of say 10°, the Q-axis would be lagging the infinite bus voltage by

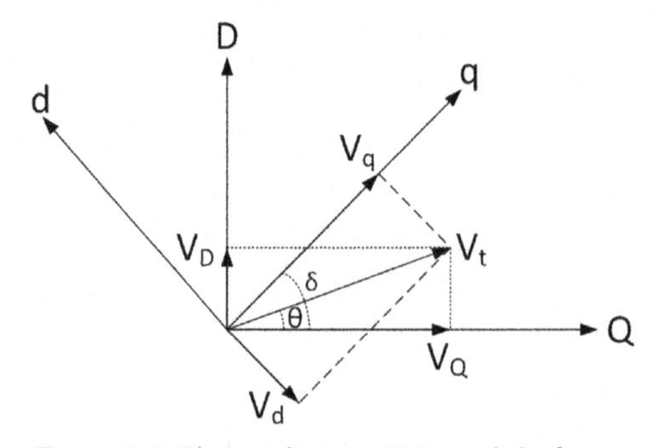

Figure 3.6 Phasor diagram DQ- and dq-frame.

10°. In a study system without an infinite bus, the angle of the slack bus and hence the network voltage reference can change during the simulation. As the slack bus angle is arbitrary, we can retrospectively change all system angles, such that the slack bus angle remains fixed along with the network voltage reference. This subtle point will become clear in Chapter 5, during discussion of the four-machine system.

θ is the voltage angle of the voltage V_t with reference to the zero degree network reference angle.

δ is the angle by which the q-axis leads the Q-axis, where the q-axis is aligned with the field flux—produced rotational EMF or speed voltage of the respective synchronous machine and the Q-axis is aligned with a network voltage angle of zero, where all network angles are measured relative to a common reference bus. All machine dq-reference frames can now be expressed relative to the DQ-reference frame.

The machine equations are in their respective dq-reference frames, which align with the field of the machine. The network equations are in the DQ-reference frame, which is a common reference frame for all machines as often imposed by the network. At the interfaces between the machines and the network, we need to convert from the dq-reference frames to the DQ-reference frame and vice versa. In Chapter 2 on power flow, we have seen that the network model takes current as an input and gives voltage as an output. The current can be, for example, from the synchronous machines into the network, and the network voltage is passed as an input to the synchronous machines, choosing the dq-reference frame, which is aligned with the respective synchronous machine. From Fig. 3.4, we can see that the conversion between dq-frame and DQ-frame can be achieved using Eqs. (3.1) and (3.2).

$$V_q + jV_d = (V_Q + jV_D)e^{-j\delta} \tag{3.1}$$

$$I_q + jI_d = (I_Q + jI_D)e^{-j\delta} \tag{3.2}$$

So far, we have talked about the dq-frame and DQ-frame; however, we have not discussed how we can convert the rotating three-phase quantities into the stationary two-phase dq-representation. Let us look once in detail at how this transformation works, later we can simply use the results. If V is the RMS line-to-line voltage, then $\sqrt{2}\ V$ is the peak value of the line-to-line voltage and V is the line-to-neutral peak voltage. In a three-phase system, the phases are equally spaced apart. With a full rotation at 2π, this means that the phases are $\frac{2\pi}{3}$ spaced apart. If ω is the rotational speed of the three phasors, we can write a vector of the three phasors. If the phasors are balanced and equally spaced, the vector is as follows:

$$V_{3ph} = \begin{bmatrix} \sqrt{\frac{2}{3}} V \sin(\omega t + \theta) \\ \sqrt{\frac{2}{3}} V \sin\left(\omega t - \frac{2\pi}{3} + \theta\right) \\ \sqrt{\frac{2}{3}} V \sin\left(\omega t + \frac{2\pi}{3} + \theta\right) \end{bmatrix} \tag{3.3}$$

The dq0 transformation is calculated by multiplying a matrix with the balanced three-phase vector, to find the dq0 vector, where the 0 element is zero.

$$V_{dq0} = \begin{bmatrix} V_d \\ V_q \\ 0 \end{bmatrix} \tag{3.4}$$

$$K * V_{3ph} = V_{dq0} \tag{3.5}$$

The inverse of this matrix is multiplied with V_{dq0} to transform the voltage back to the three-phase version.

$$K^{-1} * V_{dq0} = V_{3ph} \tag{3.6}$$

Such a matrix K has been proposed by Park and was later revised such that the transformation is power invariant.

$$K = \sqrt{\frac{2}{3}} \begin{bmatrix} \cos\gamma & \cos\left(\gamma - \frac{2\pi}{3}\right) & \cos\left(\gamma + \frac{2\pi}{3}\right) \\ \sin\gamma & \sin\left(\gamma - \frac{2\pi}{3}\right) & \sin\left(\gamma + \frac{2\pi}{3}\right) \\ \sqrt{\frac{1}{2}} & \sqrt{\frac{1}{2}} & \sqrt{\frac{1}{2}} \end{bmatrix} \tag{3.7}$$

This matrix has a very useful property when we need to find the inverse of the matrix. It is an orthogonal matrix. Let us write a simple code to check this. We chose any arbitrary angle for γ, in this example code $\frac{\pi}{8}$. Now, we calculate the value of K and the inverse of K, which is M2 and the transpose of K, which is M3. If we run Script 3.1 code, we can see that M2 and M3 are

```
gamma=pi/8;
K =sqrt(2/3)*[cos(gamma) cos(gamma -(2*pi/3)) cos(gamma
+(2*pi/3)); sin(gamma) sin(gamma -(2*pi/3)) sin(gamma
+(2*pi/3)); 1/sqrt(2)   1/sqrt(2)   1/sqrt(2)];
M2=inv(K)
M3=K.'
```

Script 3.1 Inverse and transpose of orthogonal matrix.

identical; this is because K is an orthogonal matrix that fulfils the condition $A^T * A = I$, where I is the identity matrix. Calculate M3*K and confirm if it is the identity matrix I. This result is very useful because we can take the transpose of the matrix K, rather than calculating an inverse.

$$K^{-1} = \sqrt{\frac{2}{3}} \begin{bmatrix} \cos\gamma & \sin\gamma & \sqrt{\frac{1}{2}} \\ \cos\left(\gamma - \frac{2\pi}{3}\right) & \sin\left(\gamma - \frac{2\pi}{3}\right) & \sqrt{\frac{1}{2}} \\ \cos\left(\gamma + \frac{2\pi}{3}\right) & \sin\left(\gamma + \frac{2\pi}{3}\right) & \sqrt{\frac{1}{2}} \end{bmatrix} \tag{3.8}$$

We will now calculate V_{dq0} and V_{3ph} using (3.5) and (3.6). We only need to do this once to understand the transformation; later, we can rely on the result and avoid the detailed steps.

According to (3.5), V_{dq0} is

$$\sqrt{\frac{2}{3}} \begin{bmatrix} \cos\gamma & \cos\left(\gamma - \frac{2\pi}{3}\right) & \cos\left(\gamma + \frac{2\pi}{3}\right) \\ \sin\gamma & \sin\left(\gamma - \frac{2\pi}{3}\right) & \sin\left(\gamma + \frac{2\pi}{3}\right) \\ \sqrt{\frac{1}{2}} & \sqrt{\frac{1}{2}} & \sqrt{\frac{1}{2}} \end{bmatrix} * \begin{bmatrix} \sqrt{\frac{2}{3}} V \sin(\omega t + \theta) \\ \sqrt{\frac{2}{3}} V \sin\left(\omega t - \frac{2\pi}{3} + \theta\right) \\ \sqrt{\frac{2}{3}} V \sin\left(\omega t + \frac{2\pi}{3} + \theta\right) \end{bmatrix} \tag{3.9}$$

To calculate this, we quickly need to remind ourselves of a few useful trigonometric identities.

$$\sin\alpha\cos\beta = \frac{1}{2}\left(\sin(\alpha+\beta) + \sin(\alpha-\beta)\right) \tag{3.10}$$

$$\sin\alpha\sin\beta = \frac{1}{2}\left(\cos(\alpha-\beta) - \cos(\alpha+\beta)\right) \tag{3.11}$$

$$\frac{2\pi}{3} = \frac{-4\pi}{3}, \frac{-2\pi}{3} = \frac{4\pi}{3} \tag{3.12}$$

$$\sin(-\alpha) = -\sin(\alpha) \tag{3.13}$$

$$\sin(\alpha) + \sin\left(\alpha - \frac{2\pi}{3}\right) + \sin\left(\alpha + \frac{2\pi}{3}\right) = 0 \tag{3.14}$$

We also introduce a new reference angle δ, which is rotating at the rotational speed ω such that the angle γ can be written as $\gamma = \delta + \omega t$.

First, we calculate V_d from the first row of K and V_{3ph} as seen in (3.9).

$$\begin{aligned} V_d = \frac{1}{3}V\Bigg(&\sin(\omega t + \theta + \gamma) + \sin(\omega t + \theta - \gamma) + \sin\left(\omega t + \theta + \gamma - \frac{4\pi}{3}\right) \\ &+ \sin(\omega t + \theta - \gamma) + \sin\left(\omega t + \theta + \gamma + \frac{4\pi}{3}\right) + \sin(\omega t + \theta - \gamma)\Bigg) \\ &= -V\sin(-\omega t - \theta + \gamma) = -V\sin(\delta - \theta) \end{aligned} \tag{3.15}$$

Next, we calculate V_q from the second row of K and V_{3ph} as seen in (3.9).

$$\begin{aligned} V_q = \frac{1}{3}V\Bigg(&\cos(\gamma - \omega t - \theta) - \cos(\gamma + \omega t + \theta) + \cos(\gamma - \omega t - \theta) \\ &- \cos\left(\gamma + \omega t + \theta - \frac{4\pi}{3}\right) + \cos(\gamma - \omega t - \theta) \\ &- \cos\left(\gamma + \omega t + \theta + \frac{4\pi}{3}\right)\Bigg) = V\cos(\gamma - \omega t - \theta) = V\cos(\delta - \theta) \end{aligned} \tag{3.16}$$

Finally, we calculate the zero element from the third row of K and V_{3ph} as seen in (3.9). Keeping the trigonometric identity in mind, we can see by inspection that the last element is zero. This is the complete dq0 transform. Now, let us see if we get back to the original result if we transform back to three phases. For this, we follow (3.6).

According to (3.6), V_{3ph} is

$$\sqrt{\frac{2}{3}}\begin{bmatrix} \cos\gamma & \sin\gamma & \sqrt{\frac{1}{2}} \\ \cos\left(\gamma-\frac{2\pi}{3}\right) & \sin\left(\gamma-\frac{2\pi}{3}\right) & \sqrt{\frac{1}{2}} \\ \cos\left(\gamma+\frac{2\pi}{3}\right) & \sin\left(\gamma+\frac{2\pi}{3}\right) & \sqrt{\frac{1}{2}} \end{bmatrix} * \begin{bmatrix} -V\sin(\delta-\theta) \\ V\cos(\delta-\theta) \\ 0 \end{bmatrix} \tag{3.17}$$

The equation for the first phase V_a is

$$\begin{aligned} V_a &= \sqrt{\frac{1}{6}}V(-\sin(\delta-\theta+\gamma)-\sin(\delta-\theta-\gamma)+\sin(\gamma+\delta-\theta) \\ &\quad +\sin(\gamma-\delta+\theta)) \\ &= \sqrt{\frac{1}{6}}V(-\sin(\delta-\theta-\gamma)+\sin(\gamma-\delta+\theta)) \\ &= \sqrt{\frac{2}{3}}V\sin(\gamma-\delta+\theta)=\sqrt{\frac{2}{3}}V\sin(\omega t+\theta) \end{aligned} \tag{3.18}$$

The equation for the second phase V_b is

$$\begin{aligned} V_b &= \sqrt{\frac{1}{6}}V\left(-\sin\left(\delta-\theta+\gamma-\frac{2\pi}{3}\right)-\sin\left(\delta-\theta-\gamma+\frac{2\pi}{3}\right)\right. \\ &\quad \left.+\sin\left(\gamma+\delta-\theta-\frac{2\pi}{3}\right)+\sin\left(\gamma-\delta+\theta-\frac{2\pi}{3}\right)\right) \\ &= \sqrt{\frac{2}{3}}V\sin\left(\omega t-\frac{2\pi}{3}+\theta\right) \end{aligned} \tag{3.19}$$

The equation for the third phase V_c is

$$\begin{aligned} V_c &= \sqrt{\frac{1}{6}}V\left(-\sin\left(\delta-\theta+\gamma+\frac{2\pi}{3}\right)-\sin\left(\delta-\theta-\gamma-\frac{2\pi}{3}\right)\right. \\ &\quad \left.+\sin\left(\gamma+\delta-\theta+\frac{2\pi}{3}\right)+\sin\left(\gamma+\frac{2\pi}{3}-\delta+\theta\right)\right) \\ &= \sqrt{\frac{2}{3}}V\sin\left(\omega t+\frac{2\pi}{3}+\theta\right) \end{aligned} \tag{3.20}$$

This shows that the elements V_a, V_b and V_c of V_{3ph} are still the same as before. The three rotating voltage phasors in a balanced three-phase system can hence be represented using two stationary phasors, which significantly simplify modelling. In the future, we can skip the conversion. Instead of using three time-varying phasors V_{3ph}, we work with V_t, which is $Ve^{-j(\delta-\theta)}$in dq-frame and $Ve^{j\theta}$ in the DQ-frame. We can see this using (3.15), (3.16) and (3.1).

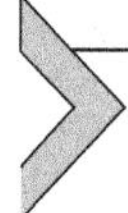

3.4 Dynamic equations of a synchronous machine in d-q reference frame

This section provides all the differential algebraic equations required for a time domain simulation of a synchronous machine. The equations capture the subtransient dynamic behaviour of a synchronous generator, with four equivalent coils as shown in Fig. 3.5 and a single lumped mass representation for the torque angle loop (Pal and Chaudhuri, 2010).

The simplest mechanical model is a single lumped mass model, as shown in Fig. 3.7. This very simple model provides information about the rotating speed of the generator, considering the applied mechanical torque and the produced electrical torque. The single-mass representation is sufficient for the power system phenomena studied in this book. The equations governing the dynamics of the single-mass model are (3.21) and (3.22).

Eq. (3.21) captures the change in generator rotor angle, due to a mismatch between the synchronous speed of the network and the rotor angular speed. $\omega_r - \omega_s$ has to be multiplied by the base value ω_b, as δ in (3.21) is an angle, which is not converted to the pu system.

$$\frac{d\delta}{dt} = \omega_b(\omega_r - \omega_s) \tag{3.21}$$

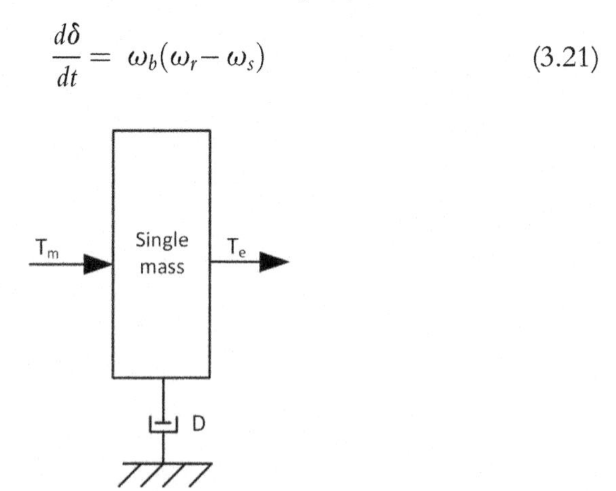

Figure 3.7 Single-mass model.

Eq. (3.22) is needed to describe the change in the rotor angular speed due to a mismatch of torques in the mechanical system. If the input mechanical torque is greater than the electrical torque, while accounting for damping, the rotor will speed up. If the electrical torque is greater than the mechanical torque, considering damping, the rotor will slow down. The rate at which the rotor speeds up or slows down is determined by the inertia of the generator.

$$\frac{d\omega_r}{dt} = \frac{1}{2H}(T_m - T_e - D(\omega_r - \omega_s)) \tag{3.22}$$

The mechanical torque is an input quantity for the synchronous machine simulation. The electrical torque can be calculated from the electrical equations of the machine, as shown in (3.23).

$$T_e = \frac{X_d'' - X_{ls}}{X_d' - X_{ls}} E_q' I_q + \frac{X_d' - X_d''}{X_d' - X_{ls}} \psi_{1d} I_q + \frac{X_d'' - X_{ls}}{X_d' - X_{ls}} E_q' I_d - \frac{X_d' - X_d''}{X_d' - X_{ls}} \psi_{2q} I_d - (X_d'' - X_d'') I_q I_d \tag{3.23}$$

The change in the transient EMF due to the field flux linkage (3.24) is indirectly proportional to the transient d-axis open-circuit time constant.

$$\frac{dE_q'}{dt} = \frac{1}{T_{d0}'}\left[-E_q' + E_{fd} + (X_d - X_d')\left(I_d + \frac{X_d' - X_d''}{(X_d' - X_{ls})^2}\left\{\psi_{1d} - E_q' - I_d(X_d' - X_{ls})\right\}\right)\right] \tag{3.24}$$

Similarly, the transient EMF due to flux linkage in q-axis damper coil (3.25) is indirectly proportional to the transient q-axis open-circuit time constant.

$$\frac{dE_d'}{dt} = \frac{1}{T_{q0}'}\left[-E_d' + \left(X_q - X_q'\right)\left(-I_q + \frac{X_q' - X_q''}{\left(X_q' - X_{ls}\right)^2}\left\{-\psi_{2q} - E_d' + I_q\left(X_q' - X_{ls}\right)\right\}\right)\right] \tag{3.25}$$

The subtransient EMF due to flux linkage in the q-axis damper coil (3.26) is indirectly proportional to the subtransient q-axis open-circuit time constant.

$$\frac{d\,\psi_{2q}}{dt} = \frac{1}{T''_{q0}}\left(-\psi_{2q} - E'_d + I_q\left(X'_q - X_{ls}\right)\right) \tag{3.26}$$

The subtransient EMF due to flux linkage in the d-axis damper coil (3.27) is indirectly proportional to the subtransient d-axis open-circuit time constant.

$$\frac{d\,\psi_{1d}}{dt} = \frac{1}{T''_{d0}}\left(-\psi_{1d} + E'_q + I_d\left(X'_d - X_{ls}\right)\right) \tag{3.27}$$

Using the four state variables and the generator terminal voltage, which is an input to the synchronous machine model, the q- and d-axis stator current components can be calculated as seen in (3.28) and (3.29), respectively.

$$\begin{aligned} I_q = & \frac{R_s}{R_s^2 + -X_d''^2}\left(E'_q\frac{X_d'' - X_{ls}}{X_d' - X_{ls}} + \psi_{1d}\frac{X_d' - X_{ls}''}{X_d' - X_{ls}} - V_q\right) \\ & + \frac{X_d''}{R_s^2 + -X_d''^2}\left(E'_d\frac{X_q'' - X_{ls}}{X_q' - X_{ls}} - \psi_{2q}\frac{X_q' - X_q''}{X_q' - X_{ls}} - V_d\right) \end{aligned} \tag{3.28}$$

$$\begin{aligned} I_d = & \frac{-X_d''}{R_s^2 + -X_d''^2}\left(E'_q\frac{X_d'' - X_{ls}}{X_d' - X_{ls}} + \psi_{1d}\frac{X_d' - X_{ls}''}{X_d' - X_{ls}} - V_q\right) \\ & + \frac{R_s}{R_s^2 + -X_d''^2}\left(E'_d\frac{X_q'' - X_{ls}}{X_q' - X_{ls}} - \psi_{2q}\frac{X_q' - X_q''}{X_q' - X_{ls}} - V_d\right) \end{aligned} \tag{3.29}$$

Synchronous machines use governors to regulate the system frequency during changes in the generation—load balance, by adjusting the input torque to the generator. A simple governor can be represented in (3.30).

$$\frac{d\,T_m}{dt} = \frac{1}{T_g}\left(T_{m2} - T_m - \frac{\omega_r - \omega_s}{R_{gov}}\right) \tag{3.30}$$

Synchronous machines use excitation systems to regulate the generator bus voltage, by adjusting the generator field voltage. A simple static excitation system can be represented in (3.31).

$$\frac{d\,E_{fd}}{dt} = \frac{1}{T_a}\left(K_a E_t - K_a V_{ref} - E_{fd}\right) \tag{3.31}$$

List of variables:

δ Generator rotor angle
ω_s Synchronous speed
$\boldsymbol{H}$ Inertia constant
$\boldsymbol{D}$ Self-damping
Transient EMF due to field flux linkage
$\boldsymbol{E_{fd}}$ Field voltage
$\boldsymbol{R_s}$ Armature resistance
ψ_{1d} Subtransient EMF due to flux linkage in d-axis damper
$\boldsymbol{I_d}$ d-axis component of stator current
V_d d-axis component of generator terminal voltage
$\boldsymbol{X_d, X'_d, X''_d}$ Synchronous, transient and subtransient d-axis reactances
$\boldsymbol{T'_{do}, T''_{do}}$ Transient and subtransient d-axis open-circuit time constants
$\boldsymbol{K_a}$ Static excitation gain
$\boldsymbol{E_t}$ Generator voltage magnitude
$\boldsymbol{T_g}$ Time constant of governor
$\boldsymbol{R_{gov}}$ Governor droop
ω_r Rotor angular speed
ω_b Base value of speed
$\boldsymbol{T_m}$ Mechanical torque
$\boldsymbol{T_e}$ Electrical torque
$\boldsymbol{E'_d}$ Transient EMF due to flux linkage in q-axis damper coil
$\boldsymbol{X_{ls}}$ Armature leakage reactance
ψ_{2q} Subtransient EMF due to flux linkage in q-axis damper
$\boldsymbol{I_q}$ q-axis component of stator current
V_q q-axis component of generator terminal voltage
$\boldsymbol{X_q, X'_q, X''_q}$ Synchronous, transient and subtransient q-axis reactances
$\boldsymbol{T'_{qo}, T''_{qo}}$ Transient and subtransient q-axis open-circuit time constants
$\boldsymbol{T_a}$ Static excitation time constant
V_{ref} Excitation voltage reference
$\boldsymbol{T_{m2}}$ Generator load reference

3.5 Initialization of the dynamic model

In this section, we examine how to find the initial conditions for the dynamic equations discussed in the last section. For this purpose, we chose a very simple study system, which we will use for the remainder of the chapter. In this system, the synchronous machine is connected via a line to an infinite bus, as shown in Fig. 3.8. This network is commonly referred to

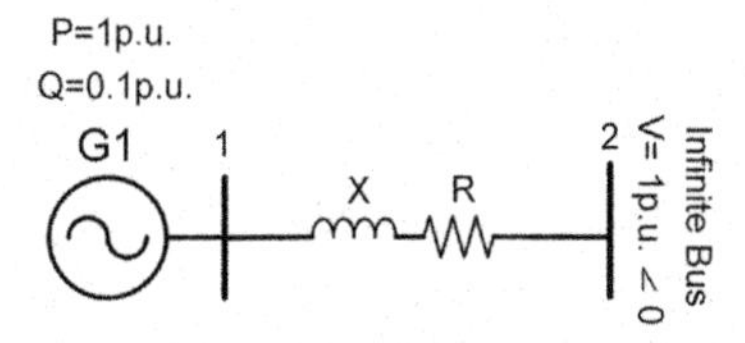

Figure 3.8 Single machine infinite bus (SMIB) system.

as SMIB system. As this will become apparent shortly, to find the initial conditions for the synchronous machine, the power flow solution, discussed in Chapter 2, is used first, to find the generator bus quantities. With the terminal quantities known, the synchronous machine initial conditions can be calculated.

An infinite bus is a bus, which has a fixed voltage (both magnitude and angle) and frequency. The infinite bus can be used to represent the connection to a strong grid, which will absorb the injected power at the infinite bus connection point, without a noticeable change in the voltage or frequency. While the voltage angle and magnitude at Bus 2 is fixed, the voltage magnitude and angle at Bus 1 varies according to the current injected at Bus 1. Please think about power system situations, where a section of a grid can be considered as a strong grid and can be replaced by an infinite bus. Can you think of an example?

The voltage at the infinite bus in this particular example is fixed as 1pu with a zero degree reference angle. Let us assume the generator has a power output of 7pu real power and 1.85pu reactive power. Normally, a typical 700 MW capacity machine will have 7.0 pu of power at 100 MW base. The power flow solution is required to find the voltage at Bus 1 and the power at Bus 2.

Start by creating a new m-file called *SMIB_data.m*. For the power flow, we require the bus and line data in the according format as described in Chapter 2. Before you read further, try and write the bus data yourself and then see if you got the same correct answer, as described in Script 3.2.

Bus 2 has a fixed voltage magnitude and angle, so it is clearly a slack bus, indicated by a 1 at the end of the line. The second and third elements for Bus 2 ensure that the bus voltage is 1pu with an angle of 0°. It was mentioned earlier that generators are often set as PV buses. In this case, this does not make sense as we know the real and reactive power and not the terminal voltage of the generator; hence, it is set to work as a PQ bus. This is indicated by a 3 at the end of the data for Bus 1.

```
% Bus No., Voltage, Angle, Pg, Qg,Pl, Ql, Gl, Bl, Bus type
bus = [...
1 1.00 0.00 7.00 1.85 0.00 0.00 0.00 0.00 3;
2 1.00 0.00 0.00 0.00 0.00 0.00 0.00 0.00 1;
];
% From bus, To bus, R L C tap ratio
line = [01 02 0.0025 0.0417 0.0000 1.0 0.0];
```

Script 3.2 Single machine infinite bus (SMIB) data.

```
% Obtain load flow solution
[Y] = form_Ymatrix(bus,line);
% calculate pre-fault power flow solution
[bus_sol, line_flow] = power_flow(Y,bus, line);
```

Script 3.3 Power flow solver.

We can use the code from Chapter 2 in Script 3.3 to find the following power flow solution:

```
bus_sol =

1 1.0520 15.8480 7.0000 1.8500 0 0 0 0 3
2 1.0000 0       -6.8816 0.1254 0 0 0 0 1

line_flow =

1 2  7.0000 1.8500
2 1 -6.8816 0.1254
```

Next, we find the initial conditions for the synchronous machine. In the introduction chapter, pu conversion and base values have been introduced, and we have used this conversion also in Chapter 2. All pu calculations for the power flow have the same base value, the system base. The parameters for synchronous machines are frequently provided in a different pu system, where the base value is the machine base, which depends on the name plate rating of the machine. When machine parameters are provided in machine base, their parameters typically vary only within a certain range. This allows power system engineers to develop an intuition of machine parameters. For example, the system base may be 100MVA, while the machine base is 900MVA. For the simulation, the engineer has two options. One is to convert the whole system to be in the pu system with the same system base. The second option is to keep the machine parameters in their individual machine base and the network in the system base and convert the parameters at the interface. In this chapter, we will introduce both methods to show that in power systems, there can sometimes be several methods that are correct in their own right. Ultimately, the results will be equivalent if both methods are correct.

First, we write our main script SMIB_run, as shown in Script 3.4, which includes the bus, line, machine, excitation and governor data. The script also solves the load flow. When executing the line *Synch_parameter_gen_base*, the machine will be initialized with the machine base, and when the first line is *Synch_parameter_sys_base*, the machine will be initialized with the system base. *generic_Sync_Init* initializes the synchronous machines. Finally, a number of system quantities are calculated to facilitate the Simulink model. Vinf is the voltage vector of the infinite bus voltage

```
%SMIB_run
clear all
system_base_mva = 100.0;
s_f=60; %System frequency in Hz
s_wb=2*pi*s_f; %Base value radial frequency in rad/sec
s_ws=1; %p.u. value of synchronous speed

% Bus No., Voltage, Angle, Pg, Qg,Pl, Ql, Gl, Bl, Bus type
bus = [...
1 1.00 0.00 7.00 1.85 0.00 0.00 0.00 0.00 3;
2 1.00 0.00 0.00 0.00 0.00 0.00 0.00 0.00 1;
];
% From bus, To bus, R L C tap ratio
line = [01 02 0.0025 0.0417 0.0000 1.0 0.0];

% MACHINE DATA STARTS

mac_con =[
1 900 0.2 0.0025 1.8 0.3 0.25 8 0.03 1.7 0.55 0.25 0.4 0.05
6.5 0];

% MACHINE DATA ENDS

% EXCITATION SYSTEM DATA
s_Ka=200;%Static excitation gain Padiyar p.328
s_Ta=0.02;%Static excitation time constant Padiyar p.328

% Governor Control SYSTEM DATA
s_Tg=0.2;%Kundur p.598
s_Rgov=0.05;%Kundur p.598

% Obtain load flow solution
[Y] = form_Ymatrix(bus,line);
% calculate pre-fault power flow solution
[bus_sol, line_flow] = power_flow(Y,bus, line);

% Synchronous machine initialisation

Smachs=[1];% buses with synchronous machines

Nbus=size(bus,1); % Number of buses
NSmachs=size(Smachs,1); % Number of synchronous machines

% program to initialise the synchronous machines

Synch_parameter_sys_base % or _gen_base, depending on the base
%system we decide to work in.

generic_Sync_Init

%Calculates the vector of apparent power S injected by the
%generators for each system bus.
PQ = bus_sol(:,4)+1i*bus_sol(:,5);
%Calculates the complex voltage value for all buses.
vol = bus_sol(:,2).*exp(1i*bus_sol(:,3)*pi/180);
%Calculates the complex value of the current injected by the
%generators at each system bus
Icalc = conj(PQ./vol);
%PL and QL are the real and reactive power drawn by loads at
%all
%all system buses.
PL = bus_sol(:,6);
QL = bus_sol(:,7);
%V is the voltage magnitude at all system buses.
V = bus_sol(:,2);

Vinfvec=ones(Nbus,1);
Vinf=Vinfvec.*(bus_sol(2,2)*exp(1i*bus_sol(2,3)*pi/180));
 %bus voltage at infinite bus
Zline=line(3)+1i*line(4); %R+jX
%Line impedance
```

Script 3.4 SMIB_run

```
%Synch_parameter_gen_base.m
machine_numb=mac_con(:,1);
machine_base_mva=mac_con(:,2);
s_xls=mac_con(:,3); %Armature leakage reactance
s_Rs=mac_con(:,4); %Armature resistance
s_xd=mac_con(:,5); %d axis synchronous reactance
s_xdd=mac_con(:,6); %d axis transient reactance
s_xddd=mac_con(:,7); %d axis sub-transient reactance
s_Tdod=mac_con(:,8); %d axis open circuit transient time
%constant
s_Tdodd=mac_con(:,9); %d axis open circuit sub-transient time
%constant
s_xq=mac_con(:,10); %q axis synchronous reactance
s_xqd=mac_con(:,11); %q axis transient reactance
s_xqdd=mac_con(:,12); %q axis sub-transient reactance
s_Tqod=mac_con(:,13); %q axis open circuit transient time
%constant
s_Tqodd=mac_con(:,14); %q axis open circuit sub-transient time
%constant
s_H= mac_con(:,15); %Inertia constant
s_D=mac_con(:,16); %Self-damping
s_Rgov = s_Rgov;

ExpandSmachs= zeros(Nbus,NSmachs);
for counter=1:NSmachs
ExpandSmachs(Smachs(counter),counter)=1;
end

SelectSmachs= zeros(NSmachs,Nbus);
for counter=1:NSmachs
SelectSmachs(counter,Smachs(counter))=1;
end

s_Pg = bus_sol(Smachs,4) .*system_base_mva./machine_base_mva;
%generator real power in machine base
s_Qg = bus_sol(Smachs,5) .*system_base_mva./machine_base_mva;
%generator reactive power in machine base
```

Script 3.5 Synch_parameter_gen_base.m in machine base.

reference in complex form for all network buses, and Zline is the impedance of the single line.

Using *Synch_parameter_gen_base*, the machine will be initialized with the machine base as shown in Script 3.5. We can see that a base conversion is required for s_Pg and s_Qg. These values are taken from the power flow results, which are in system base. If we wish to keep the machine parameters in the machine base, we need to convert the generator power also from the system base to the machine base.

Using *Synch_parameter_sys_base*, the machine will be initialized with the system base as shown in Script 3.6. The difference between both codes is highlighted in bold letters.

```
%Synch_parameter_sys_base
machine_numb=mac_con(:,1);
machine_base_mva=mac_con(:,2);
s_xls=mac_con(:,3).*system_base_mva./machine_base_mva;
%Armature leakage reactance
s_Rs=mac_con(:,4).*system_base_mva./machine_base_mva;
%Armature resistance
s_xd=mac_con(:,5).*system_base_mva./machine_base_mva; %d axis
%synchronous reactance
s_xdd=mac_con(:,6).*system_base_mva./machine_base_mva; %d axis
%transient reactance
s_xddd=mac_con(:,7).*system_base_mva./machine_base_mva; %d
%axis sub-transient reactance
s_Tdod=mac_con(:,8); %d axis open circuit transient time
%constant
s_Tdodd=mac_con(:,9); %d axis open circuit sub-transient time
%constant
s_xq=mac_con(:,10).*system_base_mva./machine_base_mva; %q axis
%synchronous reactance
s_xqd=mac_con(:,11).*system_base_mva./machine_base_mva; %q
%axis transient reactance
s_xqdd=mac_con(:,12).*system_base_mva./machine_base_mva; %q
%axis sub-transient reactance
s_Tqod=mac_con(:,13); %q axis open circuit transient time
%constant
s_Tqodd=mac_con(:,14); %q axis open circuit sub-transient time
%constant
s_H= mac_con(:,15).*machine_base_mva./system_base_mva;
%Inertia constant
s_D=mac_con(:,16).*machine_base_mva./system_base_mva; %Self-
%damping
s_Rgov = s_Rgov.*system_base_mva./machine_base_mva;

ExpandSmachs= zeros(Nbus,NSmachs);
for counter=1:NSmachs
ExpandSmachs(Smachs(counter),counter)=1;
end

SelectSmachs= zeros(NSmachs,Nbus);
for counter=1:NSmachs
SelectSmachs(counter,Smachs(counter))=1;
end

s_Pg = bus_sol(Smachs,4);
%generator real power in machine base
s_Qg = bus_sol(Smachs,5);
%generator reactive power in machine base
```

Script 3.6 Synch_parameter_sys_base.m in system base.

Once all parameters are set for initialization of the synchronous machines in machine or system base, the machines can be initialized as shown in Script 3.7 (Singh and Pal, 2013).

```
%generic_Sync_Init.m
s_f=60; %System frequency in Hz
s_wb=2*pi*s_f; %Base value radial frequency in rad/sec
s_ws=1; %p.u. value of synchronous speed

Zg= s_Rs + 1i*s_xddd; %Zg for sub-transient model
Yg=1./Zg; %Yg for sub-transient model

s_volt = bus_sol(Smachs,2); %voltage magnitude generator bus
s_theta = bus_sol(Smachs,3)*pi/180; %voltage angle in rad/sec

s_Vg = s_volt.*(cos(s_theta) + 1i*sin(s_theta));
%generator terminal voltage in complex form in DQ frame
s_Ig = conj((s_Pg+1i*s_Qg)./s_Vg);
%current at generator terminal in complex form in DQ reference
s_Eq = s_Vg + (s_Rs+1i.*s_xq).*s_Ig;
%Effective internal voltage defined as shown
s_delta = angle(s_Eq);
%Generator rotor angle

s_id = -abs(s_Ig) .* (sin(s_delta - angle(s_Ig)));
s_iq = abs(s_Ig) .* cos(s_delta - angle(s_Ig));
s_vd = -abs(s_Vg) .* (sin(s_delta - angle(s_Vg)));
s_vq = abs(s_Vg) .* cos(s_delta - angle(s_Vg));
%converting quantities from DQ-frame to dq-frame(3.1)

s_Efd = abs(s_Eq) - (s_xd-s_xq).*s_id;
%Initial value for field voltage, to follow detailed
%derivation consult (Padiyar K. R., 2008)

s_Eqd = s_Efd + (s_xd - s_xdd).* s_id;
%Derived from (3.24) and (3.27)
s_Edd = -(s_xq-s_xqd) .* s_iq;
%Derived from (3.25) and (3.26)
s_Psi1d=s_Eqd+(s_xdd-s_xls).*s_id;
%Derived from(3.27)
s_Psi2q=-s_Edd+(s_xqd-s_xls).*s_iq;
%Derived from(3.26)

s_Te = s_Eqd.*s_iq.*(s_xddd-s_xls)./(s_xdd-s_xls)+
s_Psi1d.*s_iq.*(s_xdd-s_xddd)./(s_xdd-s_xls) +
s_Edd.*s_id.*(s_xqdd-s_xls)./(s_xqd-s_xls) -
s_Psi2q.*s_id.*(s_xqd-s_xqdd)./(s_xqd-s_xls)- s_iq.*s_id.*
(s_xqdd-s_xddd);
%This is (3.23)

s_Tm=s_Te;
%Derived from (3.22)
s_iQ = cos(s_delta).*s_iq - sin(s_delta).*s_id;
s_iD = sin(s_delta).*s_iq + cos(s_delta).*s_id;
%Conversion from dq-frame to DQ-frame (3.2)

s_Vref=s_volt+1./s_Ka.*s_Efd;
%Derived from (3.31)
```

Script 3.7 generic_Sync_Init.m.

If we execute *SMIB_run* with the command *Synch_parameter_gen_base*, all required variables appear in the workspace. For example, you should be able to see that s_Tm is 0.779, s_iQ is 0.765 and s_iD is 0.014.

If we run *SMIB_run* with the command *Synch_parameter_sys_base*, s_Tm is 7.013, s_iQ is 6.881 and s_iD is 0.125.

Can you find the relationship between s_Tm in system base and machine base and check if your results make sense?

Does the same relationship hold true for s_iQ and s_iD?

3.6 Simulink modelling

To run time domain simulations for this SMIB system, we need to create a Simulink model that represents the infinite bus and line and the synchronous machine. We start by creating a new Simulink file with two subsystems, as shown in Fig. 3.9. One subsystem represents the generator and the other subsystem the network. Any changes in current injection from the generator at Bus 1 affect the network voltage, and any changes in network voltage at Bus 1 affect the generator.

The network subsystem is a representation of the equation $V_1 = V_{inf} + I_1 * Z_{line}$ and is quickly built in Simulink, as shown in Fig. 3.10. Simply double click on the network subsystem to add the components.

The generator subsystem is more complex, so we will go step by step. The generator consists of the electrical and mechanical side. The mechanical side is represented by the torque angle loop, while the electrical side is represented in the subsystem named machine. The synchronous machine also

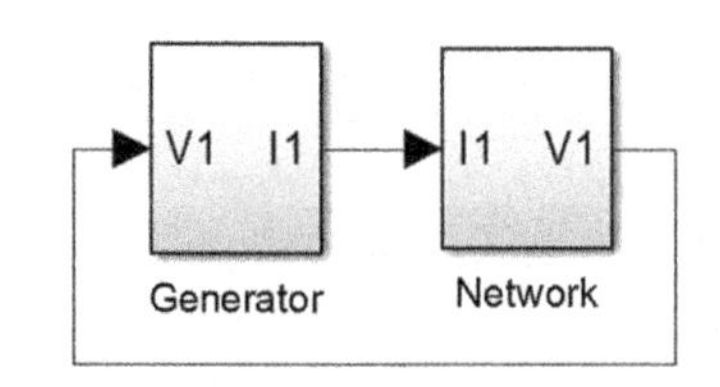

Figure 3.9 Single machine infinite bus system in Simulink.

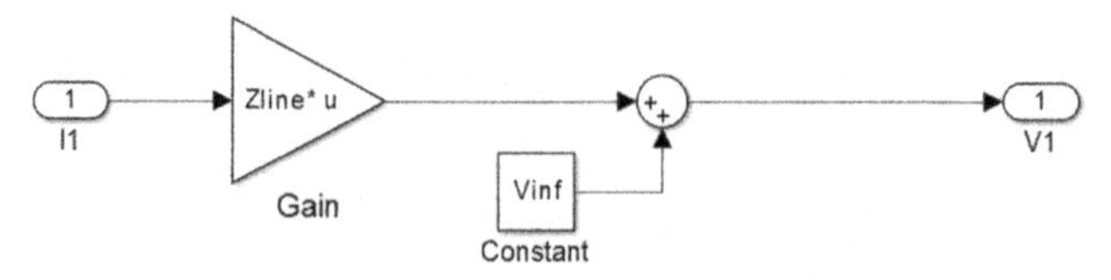

Figure 3.10 Infinite bus network in Simulink.

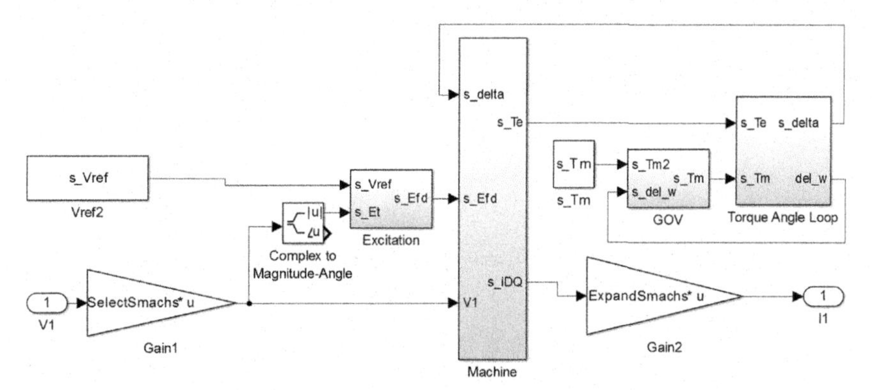

Figure 3.11 Generator representation in Simulink.

has a governor and excitation control. The gains at the voltage input and current output are required to select only the voltages of synchronous machine buses and to provide the synchronous machine currents for the synchronous machine buses. These subsystems can be seen in Fig. 3.11. Create these subsystems in your Simulink model, before we discuss what is inside of them.

The torque angle loop is built as shown in Fig. 3.12. It represents the generator mechanical dynamics given by Eqs. (3.21) and (3.22). This subsystem contains two integrator blocks, with the subtitle s_*wr* and s_*delta*. These need to be initialized. Double click on the block with subtitle s_*wr* and set the initial condition option to *s_ws*. A synchronous machine in steady state rotates at synchronous speed. For the second integrator with s_*delta*, set *s_delta* as the initial condition.

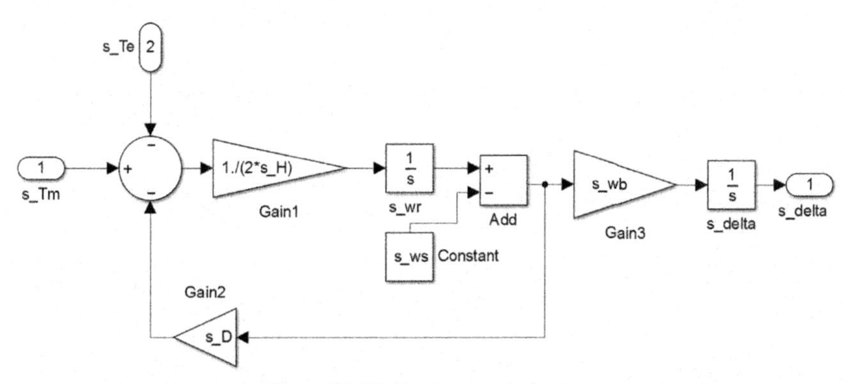

Figure 3.12 Torque angle loop.

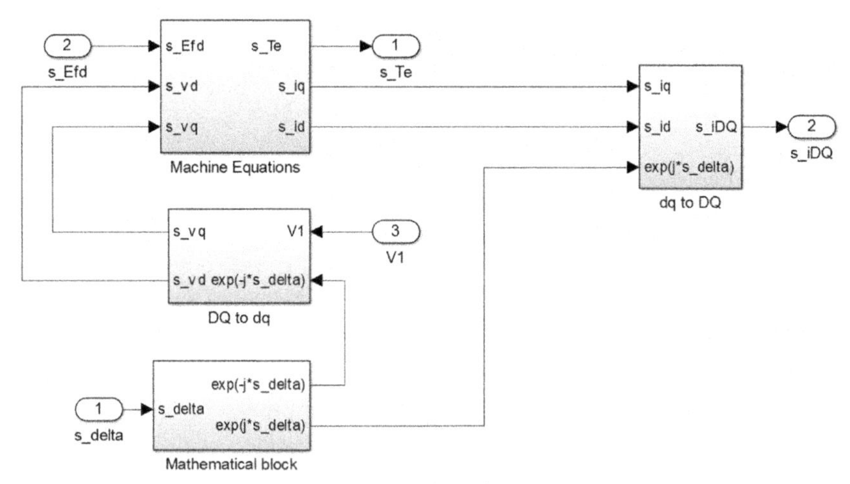

Figure 3.13 Electrical machine equations.

Fig. 3.13 shows the representation of the electrical side of the generator. We have talked about the dq-frame of generators and the DQ-frame of the network, why they are important and how to convert a quantity from one reference frame to the other. Now, we will see how to include this in our Simulink model. From the network block, the generator receives the voltage V_1 in DQ-frame, which is converted to dq-frame according to Eq. (3.1). The current that comes from the generator is in dq-frame and needs to be converted back to DQ-frame for the network according to Eq. (3.2). The machine Eqs. (3.23)−(3.29) describe what happens inside the synchronous machine.

The mathematical block creates the terms for the rotation of the phasor by $\pm\delta$. You can build it as shown in Fig. 3.14.

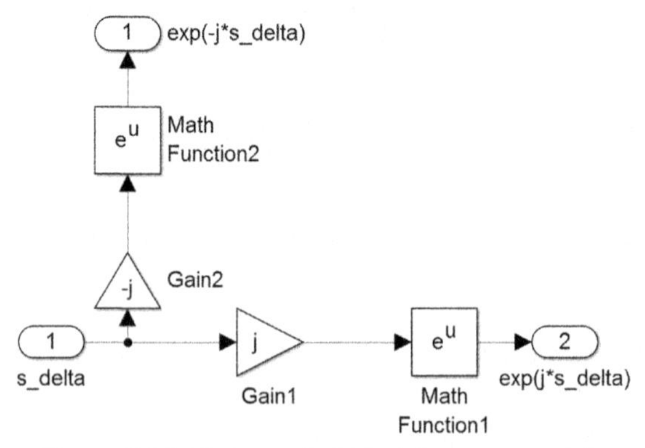

Figure 3.14 Mathematical block for exponentials.

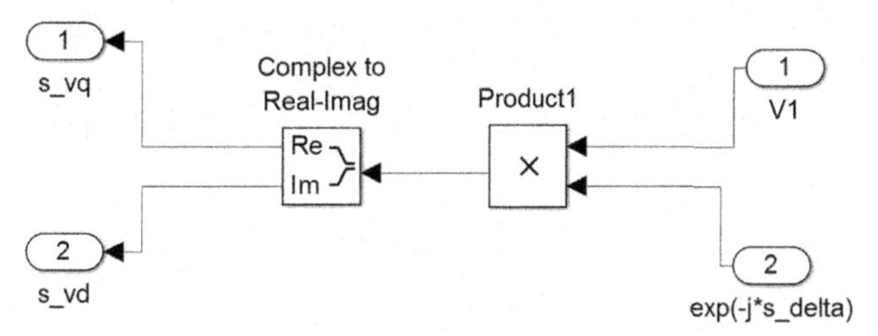

Figure 3.15 Converting from DQ- to dq-frame.

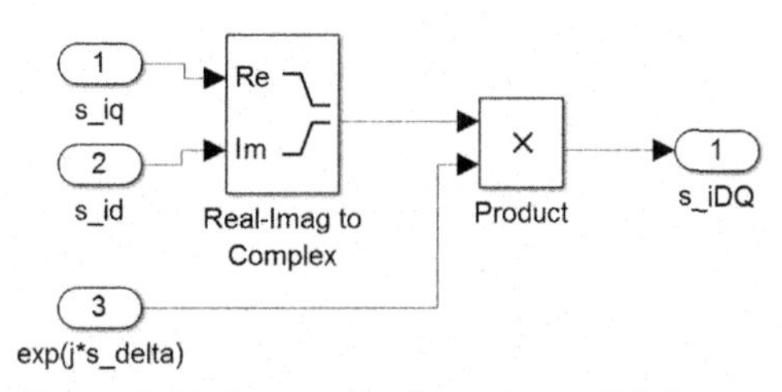

Figure 3.16 Conversion from dq- to DQ-frame.

Once the voltage is converted from DQ-frame to dq-frame, it is split into its real component (Vq) and imaginary component (Vd), as shown in Fig. 3.15.

The reverse process is used for the q- and d-current components of the generator, to convert the current to a DQ-frame for the network, as shown in Fig. 3.16.

As there are many machine equations, we use further subsystems to structure our Simulink model. This structure is shown in Fig. 3.17. Please build these subsystems before continuing to implement the individual equations.

Fig. 3.18 is a visual implementation of (3.24). Please confirm that the equation and the diagram are the same. Then implement the equation as shown here. Of course, the initial condition is s_Eqd here.

Fig. 3.19 is a visual implementation of (3.25). Implement the equation and compare the diagram with the equation. Make sure you set an appropriate initial condition here.

Can you see which equation Fig. 3.20 represents? Please ensure again to set an appropriate initial condition in the integrator block.

Please repeat the same procedure with Fig. 3.21.

Next, we need to calculate the electrical torque, as described in (3.23). The Simulink implementation of the equation can be seen in Fig. 3.22.

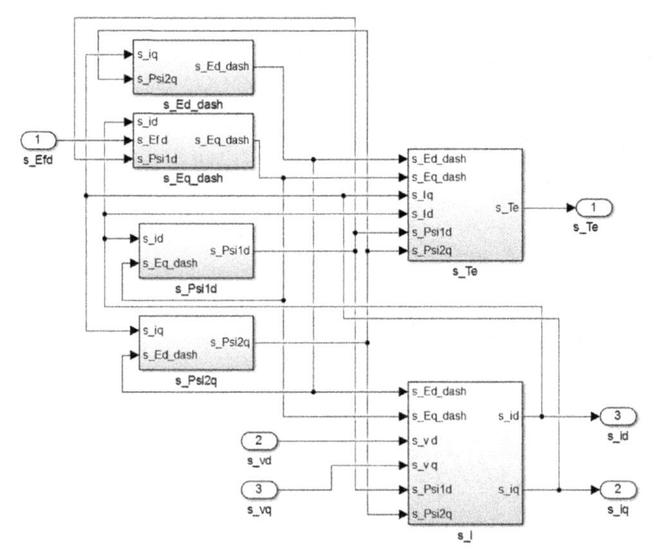

Figure 3.17 Machine equations.

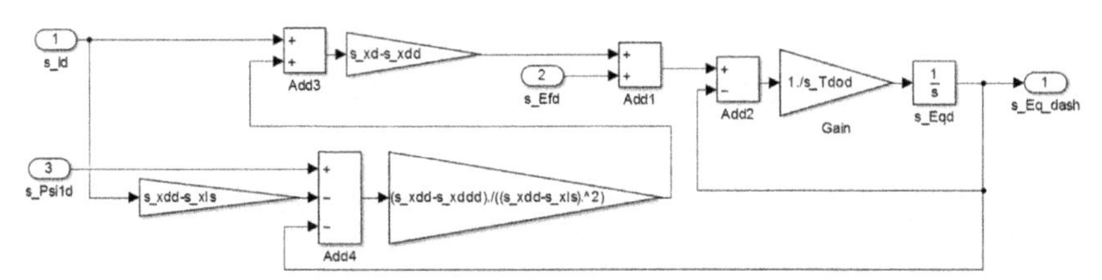

Figure 3.18 Transient EMF due to field flux linkage.

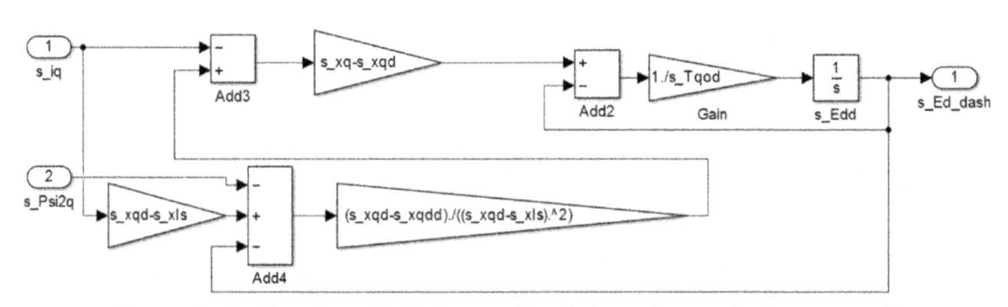

Figure 3.19 Transient EMF due to flux linkage in q-axis damper coil.

The final step is to implement Eqs. (3.28) and (3.29). As they take the same inputs, you will see in Fig. 3.23, that there is a lot of benefit implementing them as one subsystem.

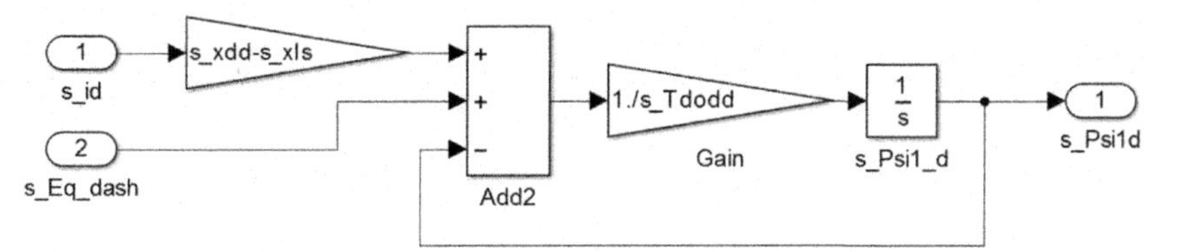

Figure 3.20 Subtransient EMF due to flux linage in d-axis damper.

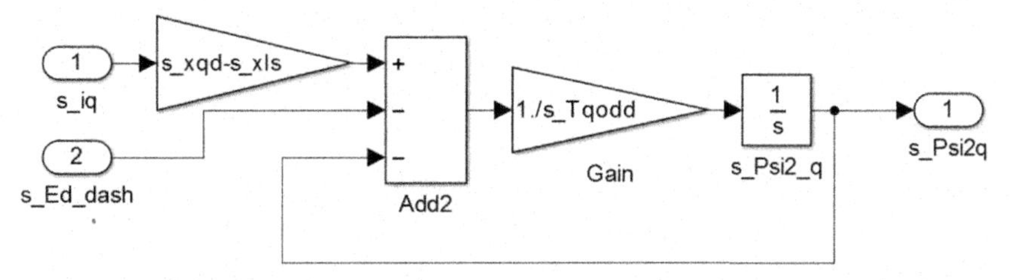

Figure 3.21 Subtransient EMF due to flux linkage in q-axis damper.

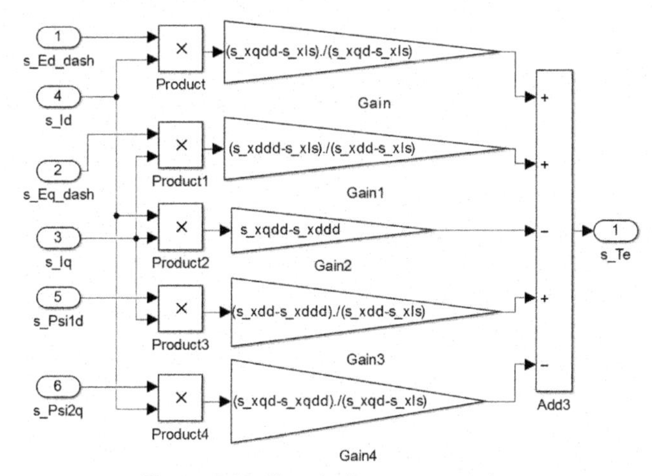

Figure 3.22 Electrical torque equation.

Can you see which equation Fig. 3.24 represents? Please ensure again to set *s_Efd./s_Ka* as initial condition in the integrator block.

Can you see which equation Fig. 3.25 represents? Please ensure again to set the correct initial condition, s_Tm.

Congratulations, you have finished building the synchronous machine model in system base. As we discussed for the initialization, you had a choice to model the machine in system base or machine base. Modelling the machine in machine base required a change of pu system at the network

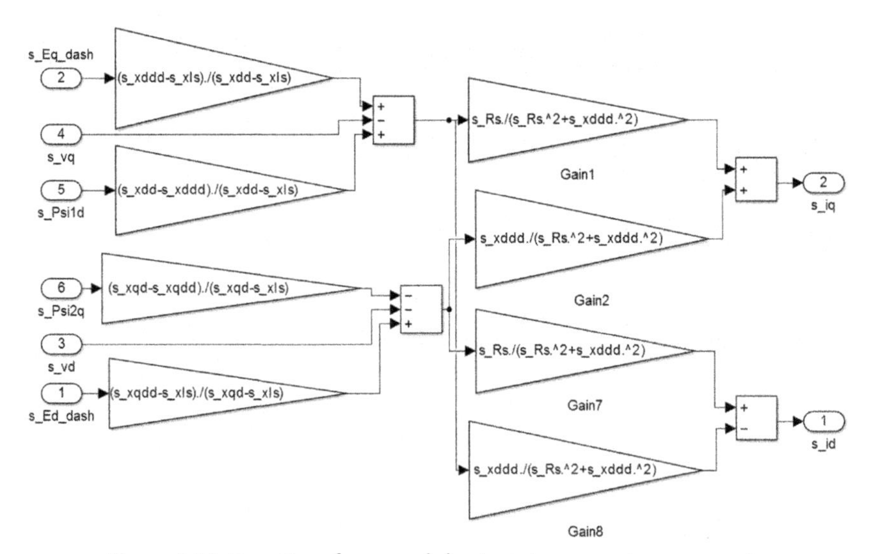

Figure 3.23 Equations for q- and d-axis stator current components.

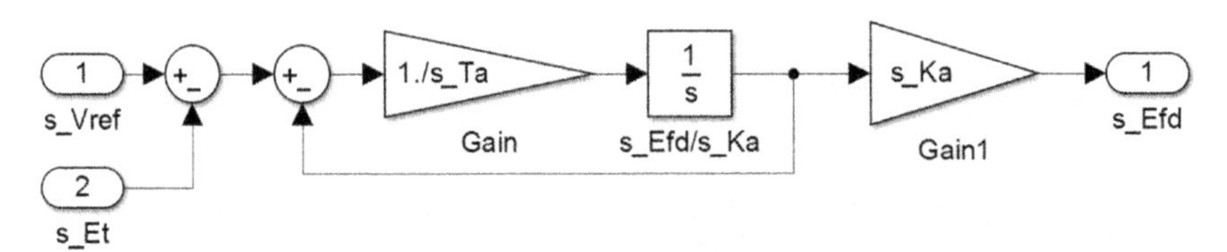

Figure 3.24 Equation for static excitation control.

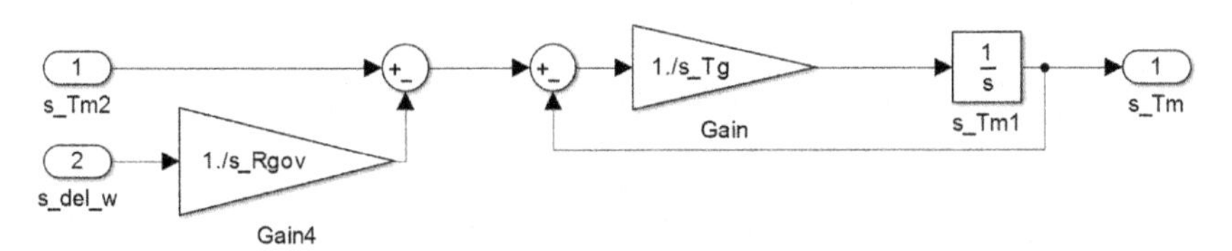

Figure 3.25 Equation for governor control.

interface. The only required change to the Simulink model is the Gain block seen in Fig. 3.26. This simulation will take the initialization code in machine base. As we can see from the display in Fig. 3.26, the electrical torque in machine base is 0.7792. Converting this to systems base, the torque is the same as in the previous simulation.

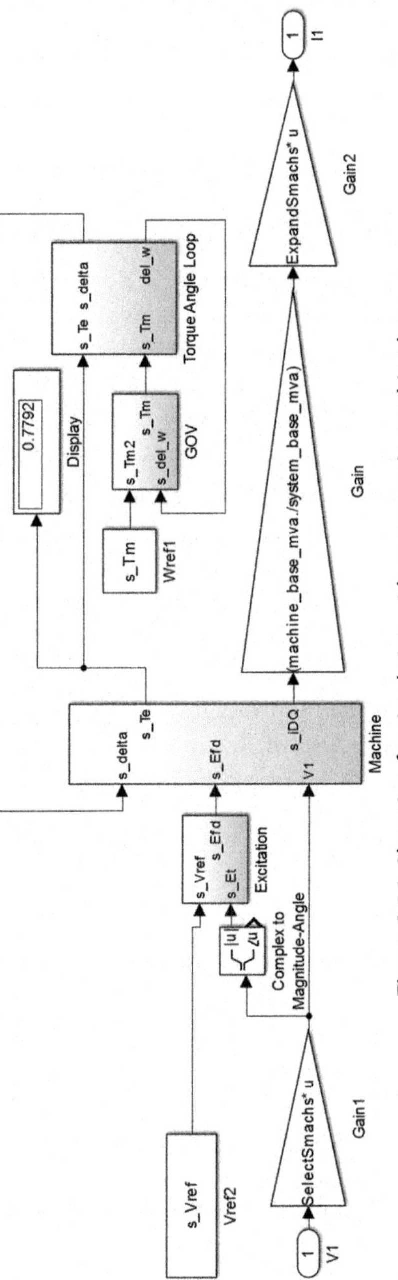

Figure 3.26 Alteration for simulation with generator in machine base.

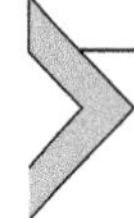

3.7 Study case: single machine infinite bus test system time domain results

Now you have implemented the whole Simulink model, all that remains is to test if it works. Please make sure you have run the Matlab initialization and all required parameters are still in the workspace. Make sure the latest changes in your Simulink model are saved and then click run. You can connect a display as a sink on the signal s_Te and run again; if you did everything 100% correct, the value shown should be 7.013.

If you do not get this value, do not worry. You will be able to gain some skills in debugging. First, see if Simulink is giving you any useful warning and messages to point to a problem. If this does not make it clear what has gone wrong, split the problem into sections. You can temporarily disconnect all the submodules, testing them a module at a time. You can start from the top, as we created them here and work your way down to individual equations. What you have to do is simple; however, it can take a bit of time. Fig. 3.27 shows an example of this approach. All inputs are replaced with constant blocks, which have the correct initial value, according to your initialization program. You can then see for each output of the subsystem, if it produces the expected output value according to what you have calculated in the initialization program. If the result differs in a subsystem, this subsystem has a problem and warrants further investigation of the equations inside, following the same procedure of disconnecting the inputs and replacing them according with the constant values, found during initialization. Continue with this until you find the problem, fix it and then make sure everything is reconnected again.

With the working model, let us run a step-response simulation. From now on, let us only use the simulation in system base. Make sure it is initialized with the according m-file. Use a step change for the mechanical input torque and use the scope and multiplexer to see how the electrical torque changes with a change in mechanical torque. At 1 second, increase the governor load reference s_Tm2 by 1% and observe the result you get (Fig. 3.28).

Fig. 3.29 shows the step response you will see, where the step change is in the mechanical input torque and the oscillatory response is seen in the electrical torque. In steady state, the mechanical torque and electrical torque are the same, which can be seen in the pre- and postdisturbance behaviour. You can also observe the change in generator rotor angle during the step change; this is shown in Fig. 3.30. Use a second display for this and change

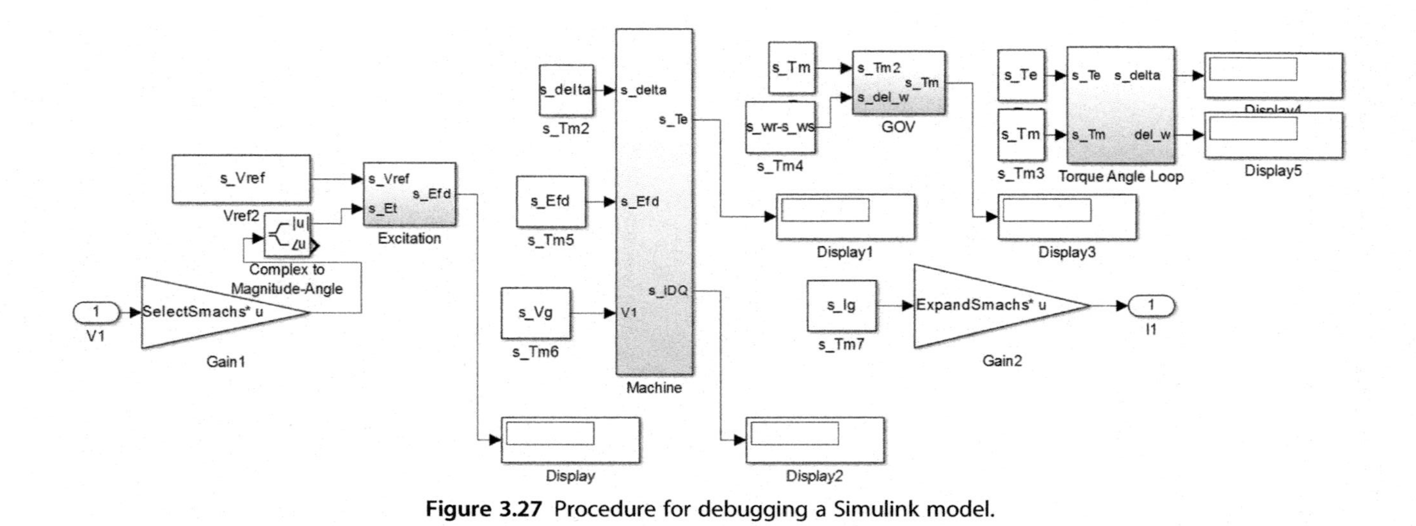

Figure 3.27 Procedure for debugging a Simulink model.

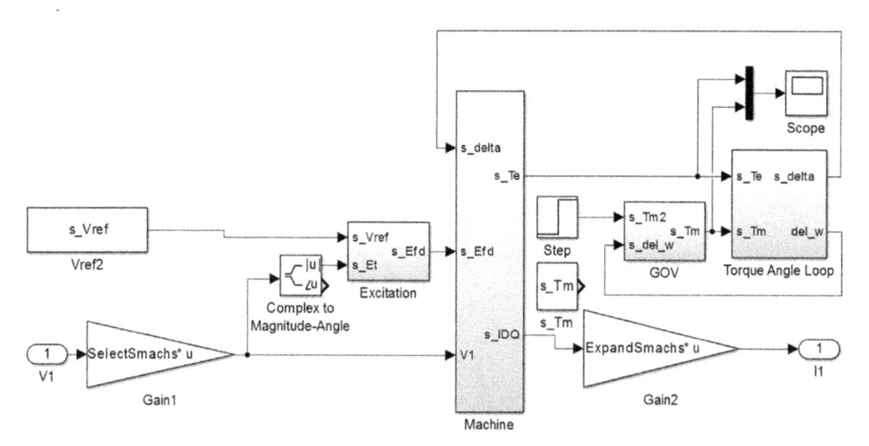

Figure 3.28 Step-response simulation.

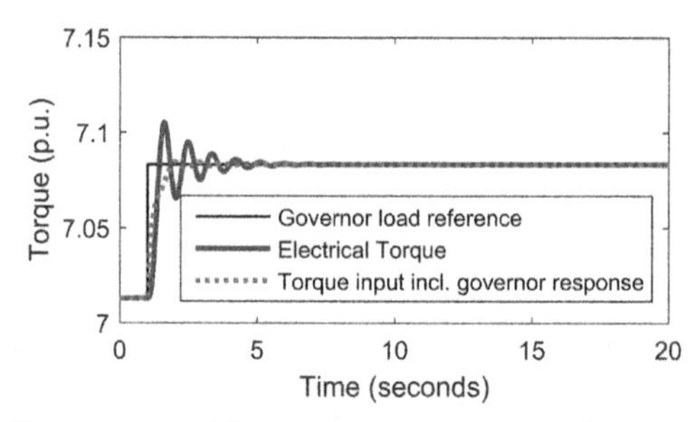

Figure 3.29 Synchronous machine step response to change in governor load reference.

the parameter setting for history, by unselecting the option limit data points to last 5000. You can then set the simulation run time to 60 s and run the simulation to see the result as shown here. Please convert the initial and final generator rotor angle into degrees.

3.8 Dynamic models of synchronous machines

The various dynamic models of synchronous machines one will encounter, working in power systems, vary in many respects. As mentioned earlier, different conventions for the dq-axis, the dq-transformation and the generator–load convention are in use.

Further differences can be found in the detail of modelling of the mechanical side. Models may range from a single-mass model to models that

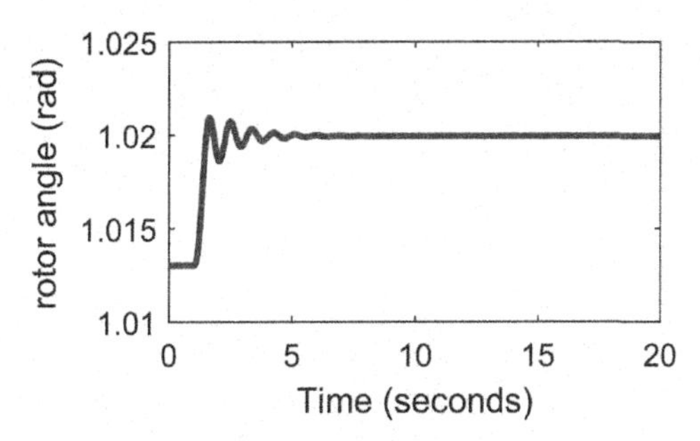

Figure 3.30 Change in generator rotor angle.

include several masses, which are coupled by springs and dampers. An example of a more detailed mechanical representation is shown in Fig. 3.31, where the mechanical side of the synchronous machine is represented by four sections, the high pressure, intermediate pressure, low pressure turbine and the generator. The diagram depicts the self-damping, mutual damping and spring constant associated with the system and the torques at various stages of the synchronous machine. The detailed mechanical model is only required for studies, where the focus is on possible interactions with the mechanical side, such as subsynchronous resonance phenomena. A detailed description of the treatment of such phenomena can be found in (Padiyar, 1999).

The representation of synchronous machines further varies according to the number of equivalent damper coils on the d- and q-axis. The number of damper coils used on each axis depends on whether the model is used to correctly capture the synchronous, transient or subtransient behaviour. Model 1.0 only has a field coil on the d-axis and no equivalent damper coils. Model 1.1 is the same as Model 1.0 with an additional equivalent damper coil on the q-axis; Model 2.1 has a further equivalent damper coil on the

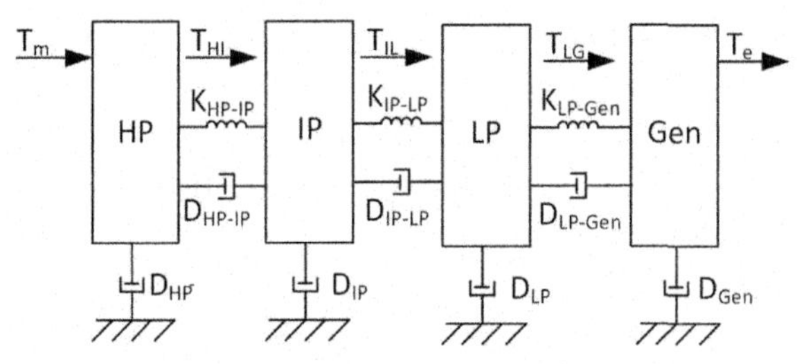

Figure 3.31 Four-mass system.

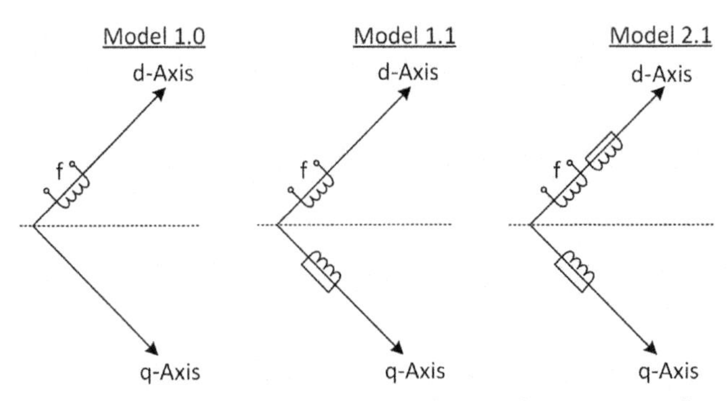

Figure 3.32 Representation of synchronous machine with varying number of equivalent damper coils.

d-axis. Model 2.2 is the one used in this book and is shown in Fig. 3.5. Padiyar, 2008 provides a detailed description of Model 1.0, 1.1, 2.1 and 2.2, which can be seen in Fig. 3.32.

We need to understand that models with a lower number of damper windings, such as Model 1.1, are only a special case of Model 2.2. For this, let us set the machine parameters as follows:

```
mac_con=[1 900 0.2 0.0025 1.8 0.3 0.25 8 0.03 1.7 0.3 0.25 0.4 0.05 6.5 0];
%Synchronous machine parameters
```

In a salient pole machine, the reactance of the quadrature axis differs from the reactance of the direct axis. Saliency can be seen in the steady-state, transient and subtransient model, where the difference is seen in the respective reactances. The modelling of saliency is described in detail by Padiyar, 2008. The model we are using does not cover saliency. By setting the parameters as shown above, we ensure that Model 1.1 has the same value as Model 2.2. Hence, there will be no saliency in either Model 1.1 or Model 2.2 in this example.

We can now run the Simulink model with the altered parameters and record the electrical torque, during a step change in mechanical torque.

Then we alter the subtransient reactances to the value of the transient reactances and rerun the simulation, with mac_con as shown below. As transient and subtransient reactances are the same, many terms of the dynamic equations drop out. This leads to Model 1.1, which has less damper windings with only synchronous and transient reactances. Please go through the generator equations and see how Model 1.1 is the special case of Model 2.2, where the transient and subtransient reactances are the same. Write out the equations of Model 1.1 by using the equations of Model 2.2 and setting X_d'' to equal X_d' and X_q'' to equal X_q'. We record the electrical torque again, during the same step change in mechanical torque as before.

```
mac_con=[1 900 0.2 0.0025 1.8 0.3 0.3 8 0.03 1.7 0.3 0.3 0.4 0.05 6.5 0];
%Synchronous machine parameters
```

Fig. 3.33 shows the result we get, if we plot the two saved electrical torque curved on top of one another. Both simulation results are identical during the steady state; however, Model 2.2 shows larger oscillations during the subtransient period, as the model includes the subtransient behaviour.

This result shows the difference between a transient model, such as Model 1.1, and a subtransient model, such as Model 2.2. We have also seen that Model 1.1 is only a special case of Model 2.2, rather than being a completely different model.

Synchronous machine models may also include magnetic saturation, experienced in actual generators. The nonlinearity introduced through saturation can be modelled by adjusting the mutual inductances for a saturation factor. A detailed description of this can be found in Kundur, 1994.

As can be seen from (3.28) and (3.29), the current components I_q and I_d are determined through a relationship of voltages V_q and V_d and the state variables of the synchronous machine. Eqs. (3.23−3.27) can be rewritten, in a more direct form, without the use of current terms I_q and I_d. The mixed current flux notation is used for ease of writing the equations. Further differences in synchronous machine models can be found, due to the timescale of the dynamics of interest. As explained in Chapter 1, the transients of a short circuit current can be split into a subtransient, transient and steady-state period. A model that aims to accurately capture the behaviour during

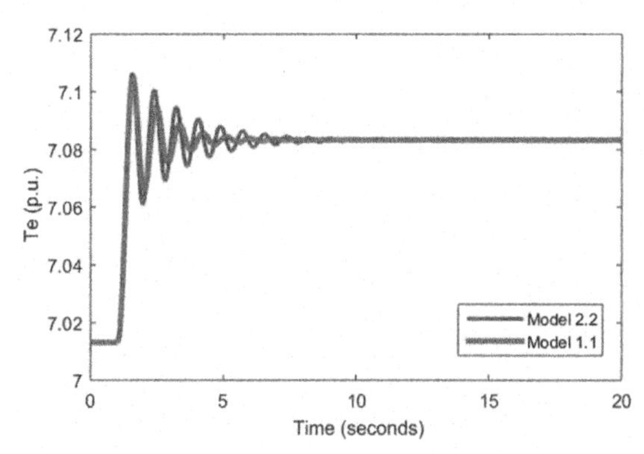

Figure 3.33 Comparison of the electrical torque Te during 1% step-up in mechanical torque using Model 2.2 and Model 1.1.

the subtransient period has to be more detailed than a model that captures the transient or steady-state period.

3.9 Simulation model of the two-area test system

The two-area test system shown in Figure 2.5 has four synchronous generators (G1, G2, G3 and G4) and two loads (at Bus-7 and Bus-9). The bus matrix and line matrix of the system are shown in Table 2.5 and Table 2.6, respectively. The synchronous machine parameters are specified using mac_con matrix given in Script 3.8. The script also lists the gain and

```
% COPY BUS DATA FROM TABLE 2.5
% COPY LINE DATA FROM TABLE 2.6

% ********************** MACHINE DATA STARTS **********************

% Machine data format
%        1. machine number,
%        2. bus number,
%        3. base mva,
%        4. leakage reactance x_l(pu),
%        5. resistance r_a(pu),
%        6. d-axis sychronous reactance x_d(pu),
%        7. d-axis transient reactance x'_d(pu),
%        8. d-axis subtransient reactance x"_d(pu),
%        9. d-axis open-circuit time constant T'_do(sec),
%       10. d-axis open-circuit subtransient time constant
%                 T"_do(sec),
%       11. q-axis sychronous reactance x_q(pu),
%       12. q-axis transient reactance x'_q(pu),
%       13. q-axis subtransient reactance x"_q(pu),
%       14. q-axis open-circuit time constant T'_qo(sec),
%       15. q-axis open circuit subtransient time constant
%                 T"_qo(sec),
%       16. inertia constant H(sec),
%       17. damping coefficient d_o(pu),
%       18. dampling coefficient d_1(pu),
%       19. bus number
%       20. saturation factor S(1.0)
%       21. saturation factor S(1.2)
% note: all the following machines use subtransient reactance model

mac_con =[
1 900 0.2 0.0025 1.8 0.3 0.25 8 0.03 1.7 0.55 0.25 0.4 0.05 6.5 0;
 2 900 0.2 0.0025 1.8 0.3 0.25 8 0.03 1.7 0.55 0.25 0.4 0.05 6.5 0;
 3 900 0.2 0.0025 1.8 0.3 0.25 8 0.03 1.7 0.55 0.25 0.4 0.05 6.5 0;
 4 900 0.2 0.0025 1.8 0.3 0.25 8 0.03 1.7 0.55 0.25 0.4 0.05 6.5 0];
% ********************** MACHINE DATA ENDS **********************

% ********************** EXCITATION SYSTEM DATA *****************
s_Ka=200;%Static excitation gain Padiyar p.328
s_Ta=0.02;%Static excitation time constant Padiyar p.328

% **** Governor Control SYSTEM DATA - not used ********
s_Tg=0.2;%Kundur p.598
s_Rgov=0.05;%Kundur p.598
```

Script 3.8 Two-area test system data.

```
% Save this program as find_impedance_matrix.m
% Copy script from Script 2.5 here

% Finding post-fault impedance matrix.
% Define post-fault admittance matrix
Yf = Y;
% Apply three phase fault at bus-8 by specifying very large
%admittance at Yf(8,8)
Yf(8,8) = 10000;
% Find post-fault impedance matrix.
Zf = inv(Yf);
```

Script 3.9 Program to find prefault and postfault impedance.

time constant of the static excitation system. Combine Table 2.5 and Table 2.6 as mentioned in Script 3.8, and save it as four_mac_data.m.

We will reuse programs and models developed so far to build a simulation model for the two-area test system. Please ensure the functions form_Ymatrix(bs,ln) and power_flow(Y, bs, ln) given in Section 2.6 are saved as form_Ymatrix.m and power_flow.m, respectively, in the current Matlab working folder. Also save Synch_parameter_sys_base.m and generic_Sync_Init.m in the current working folder.

In this chapter, we will simulate a three-phase ground fault at Bus-8 of the two-area test system. To simulate a three-phase ground fault, prefault and postfault impedance matrices are required. A three-phase ground fault at Bus-8 means a zero impedance from the bus to ground at the bus. Alternatively, we can add a large admittance at the diagonal entry corresponding to the bus in the admittance matrix and find the inverse of the new admittance matrix to get the postfault impedance matrix. Create program find_impedance_matrix.m using Script 3.9.

3.9.1 Simulink block representing multiple synchronous machines

In Section 3.5, we developed Simulink block for a synchronous machine. But the two-area test system has four synchronous machines in it. The synchronous machine Simulink block could be directly used to simulate any number of machines as long as the program is made to handle vector calculation. For example, the equation $A = B + CD$ can be represented in Matlab using `A = B + C*D` or `A = B + C.*D`. Both representations are correct. But the second one can be used for both scalar and vector calculation, as the '.' (dot) symbol before the * specifies dot product of two vectors. Hence, if second command is used, the size of vector A can be one or many. All figures in Section 3.5 show vector calculations. Readers encouraged to double check their model, as this is one of the common mistakes in early programming of multimachine system. For results presented in this section and also in Chapter 9, it is assumed that governors are not active in synchronous machines. Disconnect Governor model by connecting s_Tm constant block in Fig. 3.28 to s_Tm input of the torque angle loop block. By doing this, mechanical input torque to the generators will be constant during the course of the simulation. The change is made for ease of illustration of interarea oscillation mode, and the reader is encouraged to study the effect of governor on dynamics of the system.

Another important aspect is ensuring that correct number of inputs is passed to each block. Fig. 3.34 shows the Simulink program for the two-area system, where SelectSmachs and ExpandSmachs gains are shown outside the Generator block for clarity. The Generator block accepts a vector of generator bus voltages as input. Size of this vector must equal the number of generators present in the system. So correct bus voltages must be selected from a vector of all bus voltage outputs from the network block. There are different ways to do this. In Fig. 3.34, a 4x11 matrix, 'SelectSmachs', multiplies the bus voltage vector to obtain 4x1 generator bus voltages. Similarly, the network block requires a vector of bus current injections at all buses. In our example, the Generator block produces a current vector of size 4 corresponding to number of synchronous machines. The 'ExpandSmachs' block transforms this vector to a new vector of size 11 equal to the number of buses in the system. In the new vector, the output currents of synchronous generators are placed at the position corresponding to the synchronous machine buses and all other elements are set to zero. Script for both the

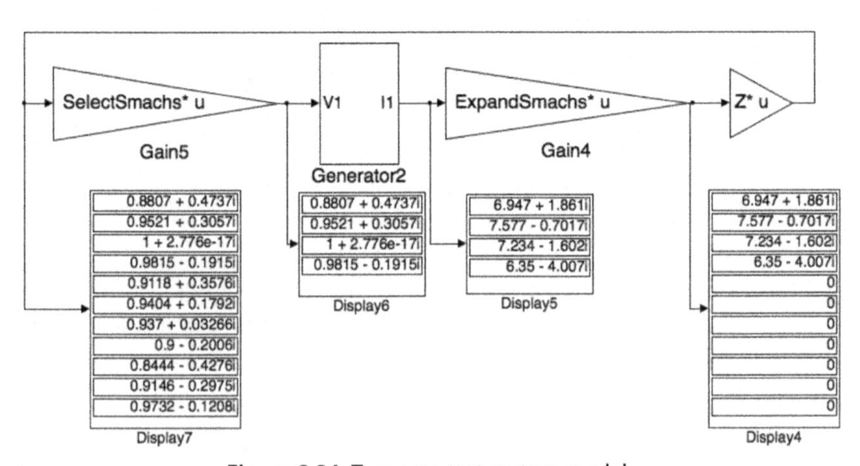

Figure 3.34 Two-area test system model.

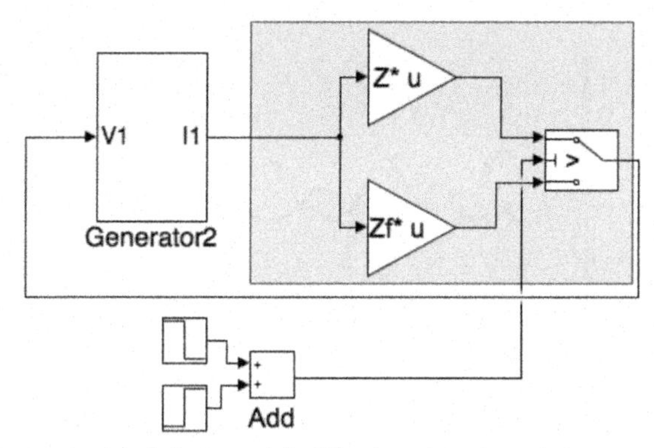

Figure 3.35 Dynamic simulation model of the two-area test system model to simulate three-phase ground fault.

matrices is given in Script 3.6. Now, build a Simulink model as shown in Fig. 3.35 using the synchronous machine block developed in Section 3.5.

Few more lines of code are required before we can run the model. Create two_area_synch.m using the code and instruction in Script 3.10. Build a Simulink model, two_area_synch_model.slx as shown in Fig. 3.35. The model simulates a three-phase ground fault at Bus-8 from

```
clear all

system_base_mva = 100.0;

% Copy the code in the Script 3.8

% Obtain load flow solution
[Y] = form_Ymatrix(bus,line);
% calculate pre-fault power flow solution
[bus_sol, line_flow] = power_flow(Y,bus, line);

% Synchronous machine initialisation

Smachs=[1;2;3;4];% buses with synchronous machines
Nbus=size(bus,1); % Number of buses
NSmachs=size(Smachs,1); % Number of synchronous machines


% program to initialise the synchronous machines
Synch_parameter_sys_base_fourmach
generic_Sync_Init

% Copy the code in the Script 3.9
```

Script 3.10 Program to initialize two-area test system model.

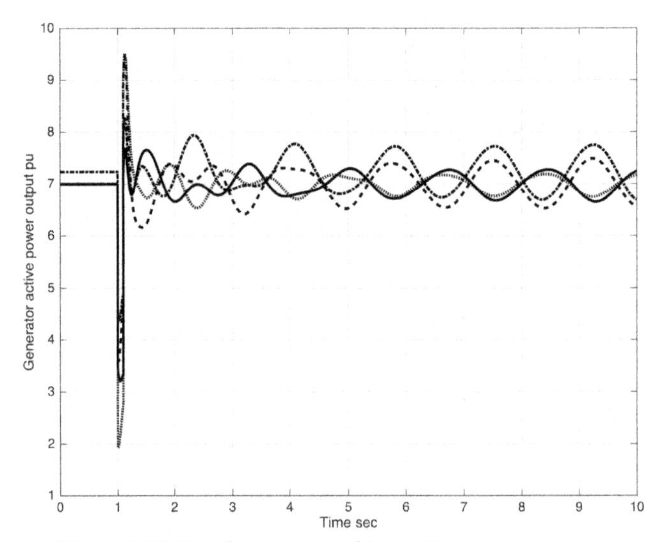

Figure 3.36 Synchronous machine active power output.

1 s to 1.1 s. This is achieved using the two gains having impedance matrices (Z and Zf) and a switch. The threshold parameter of the switch is set to 0.5. The switch selects the path of $Z{*}u$ (first input) when the second input is less than 0.5 (during normal operating condition) and changes to $Zf{*}\ u$ when second input is more than 0.5 (between time 1 and 1.1 s). Two step signals and an add block create the second input to the switch. One of the step signals changes from 1 to 0 at time 1.0 s and another step signal changes from 0 to 1 at time 1.1 s.

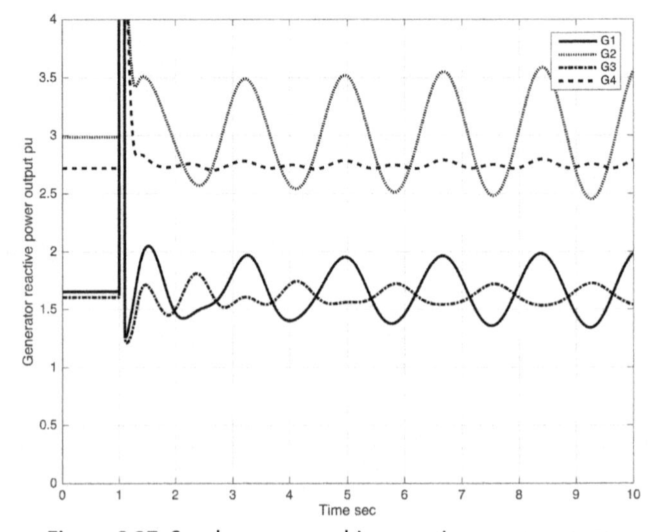

Figure 3.37 Synchronous machine reactive power output.

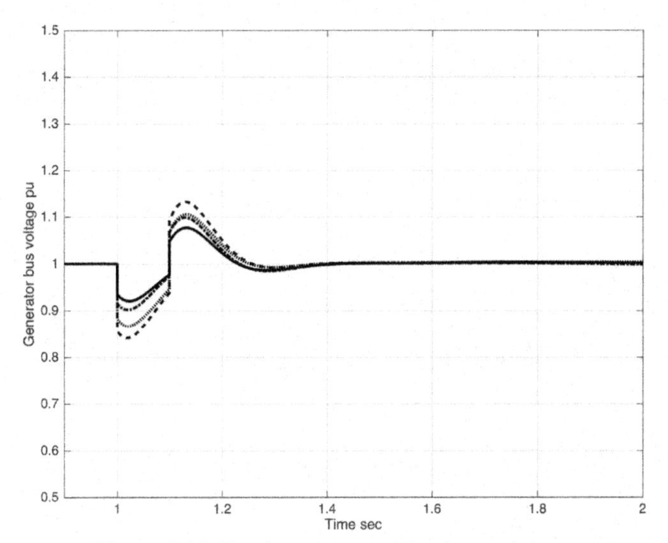

Figure 3.38 Synchronous machine bus voltages.

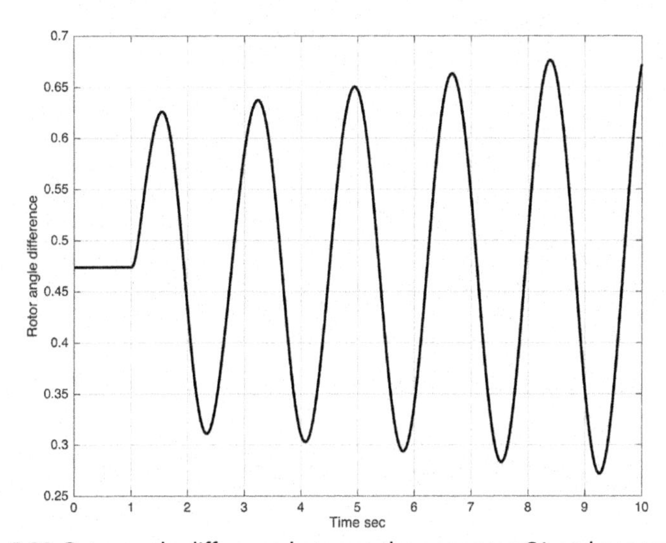

Figure 3.39 Rotor angle difference between the generator G1 and generator G3.

Now you are ready to test your model. Use sample results in Figs. 3.36–3.39 to validate your results. The plots show generator active power outputs, reactive power outputs, terminal voltages and angle difference between G1 and G3.

References

Clarke, E., 1957. Circuit Analysis of A-C Power Systems. John Wiley & Sons, Inc, New York.

IEEE committee report, 1969. Recommended phasor diagram for synchronous machines. IEEE Trans. Power Apparatus Syst. 1593–1610.

Kundur, P., 1994. Power System Stability and Control. McGraw-Hill.

Padiyar, K., 1999. Analysis of Subsynchronous Resonance in Power Systems. Kluwer Academic Publishers, Norwell.

Padiyar, K.R., 2008. Power System Dynamics Stability and Control, second ed. BS Publications, Hyderabad.

Pal, B., Chaudhuri, B., 2010. Robust Control in Power Systems. Springer, New York.

Singh, A.K., Pal, B.C., 2013. IEEE PES Task Force on Benchmark Systems for Stability Controls, Report on the 68-Bus, 16-Machine, 5-Area System. Version 3.3.

Neidhofer, G., October 1992. The evolution of the synchronous machine. Engineering Science and Education Journal 239–248.

CHAPTER FOUR

Analysis and controller design ideas

This chapter discusses various control system tools available in Matlab for stability analysis and controller design. The model of synchronous machine developed in Chapter 3 is used for the discussion. Connection between location of poles and frequency and damping ratio of dynamic response is presented in relation with eigenvalue and Bode plot. Worked out examples of stability analysis and controller design for model power system are provided. By the end of this chapter, we expect the readers to acquire required understanding to investigate further into different stability analysis and control techniques.

4.1 System representations and dynamic response

The power system model developed in Chapter 3 is nonlinear. Engineers often linearize these models around an operating condition for dynamic stability analysis and controller design. In this section, we will define some of the important terms in the linear system analysis such as eigenvalues, eigenvectors, participation factors, transfer function, poles, zeros and residue. The application and understanding of these tools and terms in power system context will be discussed in the following section.

Equation (4.1) shows a state space representation of a linear model.

$$\begin{aligned} \dot{\Delta x} &= A\,\Delta x + B\,\Delta u \\ \Delta y &= C\,\Delta x + D\,\Delta u \end{aligned} \tag{4.1}$$

where x, u and y are state, input and output vectors, respectively, and A, B, C and D are state matrix, input matrix, output matrix and feed forward matrix, respectively. Most often D is zero.

Eigenvalues and eigenvectors of matrix A describe stability of the linear system. Eigenvalues are values of vector λ such that $A\phi = \lambda\phi$, where $\phi \neq 0$. Eigenvalues are obtained from the solution of the characteristics equation given by $\det(A - \lambda I) = 0$. Eigenvalues can be real or complex; when complex, they occur in conjugate pairs of the form $\lambda_i = \sigma_i \pm \omega_i$.

Simulation of Power System with Renewables
ISBN: 978-0-12-811187-1
https://doi.org/10.1016/B978-0-12-811187-1.00004-4
© 2020 Elsevier Inc.
All rights reserved.

A right eigenvector ϕ_i corresponding to λ_i is given by $A\phi_i = \lambda_i\phi_i$. Similarly, a left eigenvector ψ_i corresponding to λ_i is given by $\psi_i = \lambda_i\psi_i$. Once ϕ_i and ψ_i corresponding to a mode λ_i of a state matrix A are found, participation factor vector PF_i can be defined as

$$PF_i = \begin{bmatrix} PF_{1i} \\ PF_{2i} \\ \vdots \\ PF_{ni} \end{bmatrix} = \begin{bmatrix} \phi_{1i}\psi_{i1} \\ \phi_{2i}\psi_{i2} \\ \vdots \\ \phi_{ni}\psi_{in} \end{bmatrix} \tag{4.2}$$

where ϕ_{ki} and ψ_{ik} are kth elements of the vectors ϕ_i and ψ_i, respectively. The participation factor indicates relative participation of kth state x_k on ith mode λ_i. We will explore this using an example later in this chapter.

The state space representation in (4.1) can be converted to an input—output transfer function as

$$G(s) = \frac{\Delta y}{\Delta x} = C(sI - A)^{-1}B + D \tag{4.3}$$

$$G(s) = K\frac{(s - z_i)(s - z_2)\cdots(s - z_m)}{(s - p_1)(s - p_2)\cdots(s - p_n)} \tag{4.4}$$

where K is a constant, $p_1,\ p_2,\ \ldots p_n$ are poles of $G(s)$ and $z_1, z_2 \ldots. z_m$ are zeros of $G(s)$. The poles are same as eigenvalues of A.

Another form of $G(s)$ is using residue-pole representation as

$$G(s) = \frac{R_1}{s - \lambda_1} + \frac{R_2}{s - \lambda_2} + \cdots\frac{R_n}{s - \lambda_n} \tag{4.5}$$

where R_i is residue of $G(s)$ at pole λ_i and

$$R_i = C\phi_i\psi_iB \tag{4.6}$$

Residue is a very useful quantity in controller design. To appreciate the residue, it is important to know observability and controllability. Observability indicates how well an oscillation is visible in an output, whereas controllability indicates how effectively an input influences an oscillatory mode. This is a very simplistic definition. Residue is the product of the observability and controllability, which means if R_i corresponding to λ_i is large in a transfer function, a feedback controller between the corresponding input and output can effectively change location of λ_i. Magnitude of residue is a good indicator to select input actuator and output signal for controller design, when several input—output combinations are available.

4.1.1 Stability of the linear system

Stability of a linear time invariant system depends on position of its eigenvalues on a complex plane. If any eigenvalues of a system are on the right half of the plane, meaning eigenvalues with positive real part, the system is unstable. A system is said to be marginally stable if eigenvalues are present on the imaginary axis. A real negative eigenvalue results in stable decaying response. The larger the magnitude, the faster the decay. Complex eigenvalues, which occur as conjugate pairs, represent oscillatory behaviour. A complex eigenvalue with negative real part results in a damped oscillation, whereas one with positive real part results in an oscillation of increasing amplitude. The real part in this case represents damping, and imaginary component represents frequency of oscillation. For an eigenvalue, $\lambda = \sigma \pm j\omega$, damping ratio $\zeta = -100*\frac{\sigma}{|\lambda|}\%$ and frequency $f = \frac{\omega}{2\pi}$ Hz. The time constant of amplitude decay depends on the real part of eigenvalue, which is $\frac{1}{|\sigma|}$. This means the amplitude decays to $1/e$ or 37% of initial value in $\frac{1}{|\sigma|}$ seconds or $\frac{1}{2\pi\zeta}$ cycles.

4.1.1.1 Exercise 4.1

Using the terms defined so far, let us visualize the relationship between eigenvalues and system response through a set of 12 transfer functions listed in Table 4.1. Figs. 4.1–4.12 show eigenvalue plot, Bode plot and step response plots corresponding to the transfer functions. Eigen plot shows location of the poles on a complex plain. Bode plot shows gain and phase

Table 4.1 Example transfer functions.

Name	Frequency (Hz)	Damping ratio (%)	Transfer function $G(s)$
T1	10	5	$\frac{s^2}{s^2+6.291s+3958}$
T2	10	1	$\frac{s^2}{s^2+1.257s+3948}$
T3	10	0	$\frac{s^2}{s^2+3948}$
T4	10	−1	$\frac{s^2}{s^2-1.257s+3948}$
T5	10	−5	$\frac{s^2}{s^2-6.291s+3958}$
T6	100	5	$\frac{s^2}{s^2+62.91s+395800}$
T7	100	1	$\frac{s^2}{s^2+12.57s+394800}$
T8	100	0	$\frac{s^2}{s^2+394800}$
T9	100	−1	$\frac{s^2}{s^2-12.57s+394800}$
T10	100	−5	$\frac{s^2}{s^2-62.91s+395800}$
T11	0	100	$\frac{1}{s+20}$
T12	0	−100	$\frac{1}{s-20}$

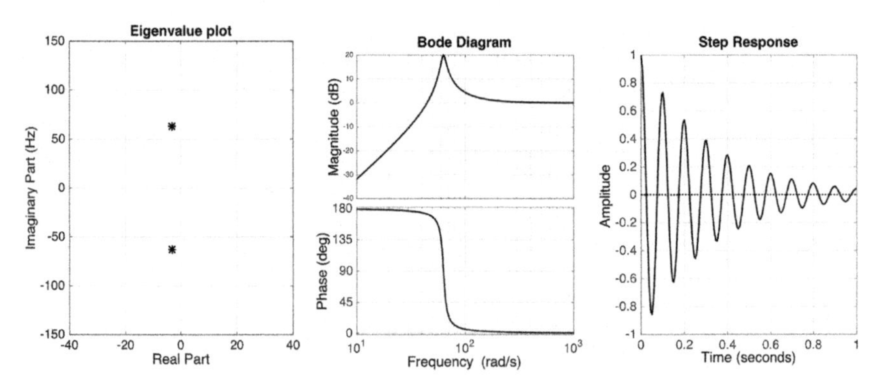

Figure 4.1 Eigenvalue, Bode and dynamic response plots of T1.

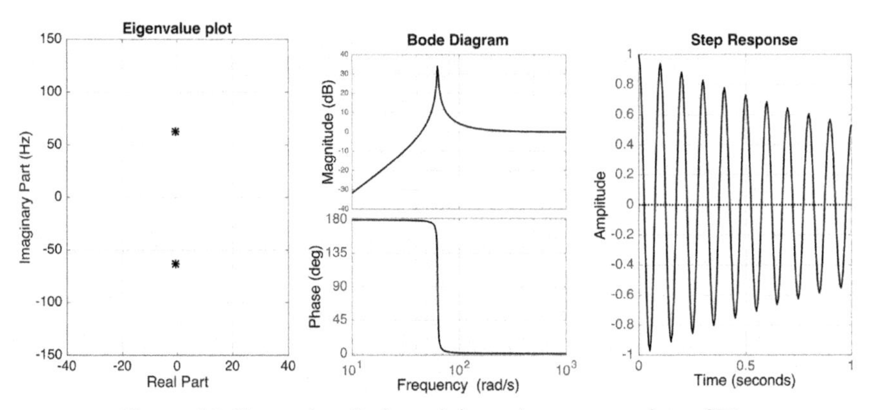

Figure 4.2 Eigenvalue, Bode and dynamic response plots of T2.

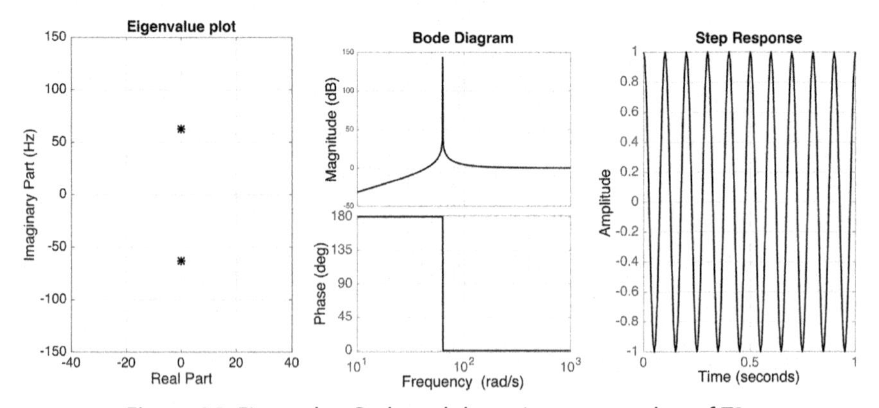

Figure 4.3 Eigenvalue, Bode and dynamic response plots of T3.

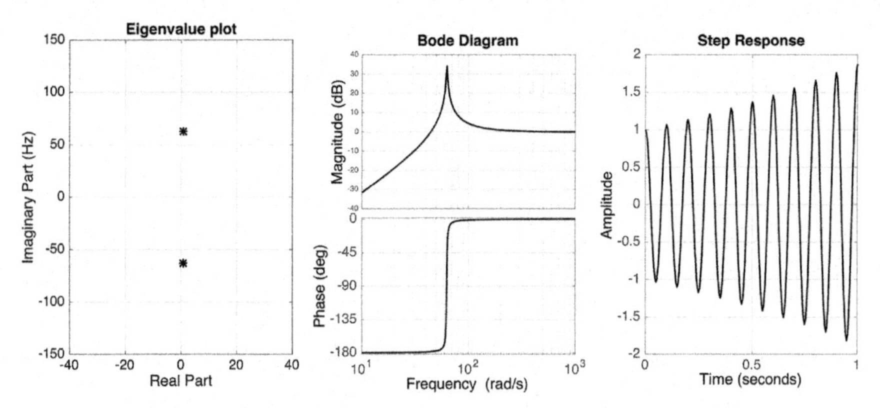

Figure 4.4 Eigenvalue, Bode and dynamic response plots of T4.

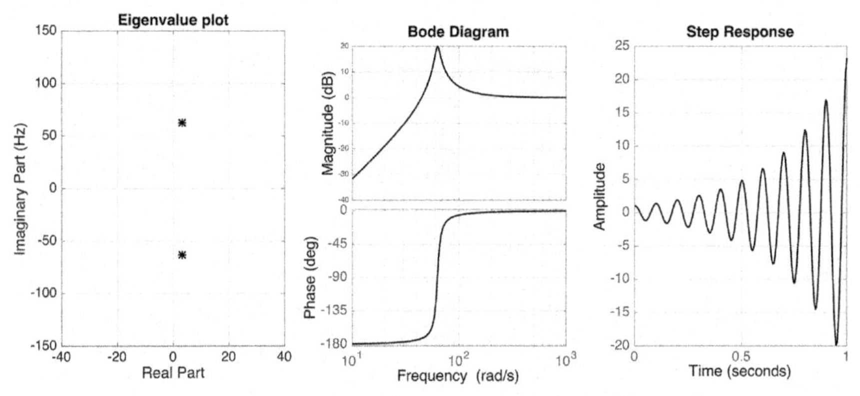

Figure 4.5 Eigenvalue, Bode and dynamic response plots of T5.

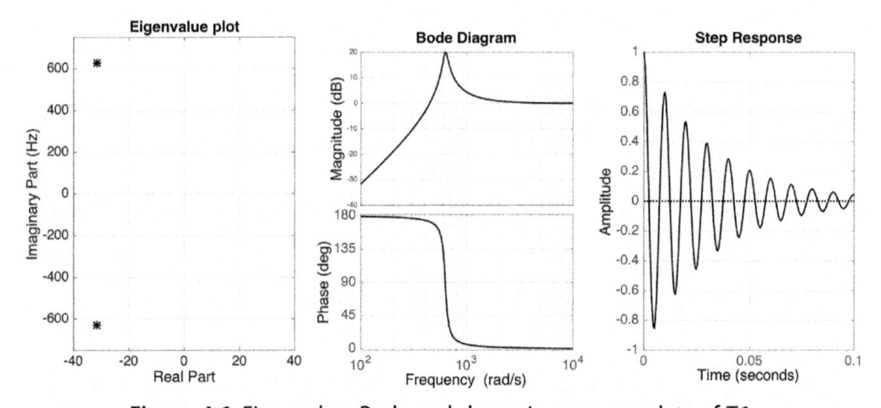

Figure 4.6 Eigenvalue, Bode and dynamic response plots of T6.

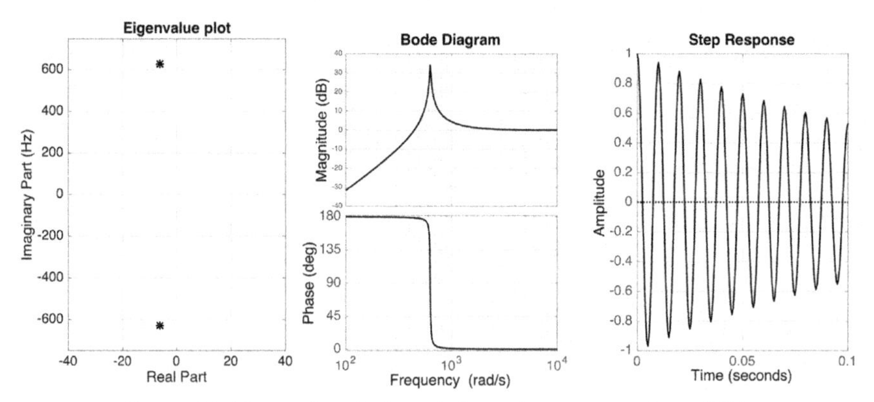

Figure 4.7 Eigenvalue, Bode and dynamic response plots of T7.

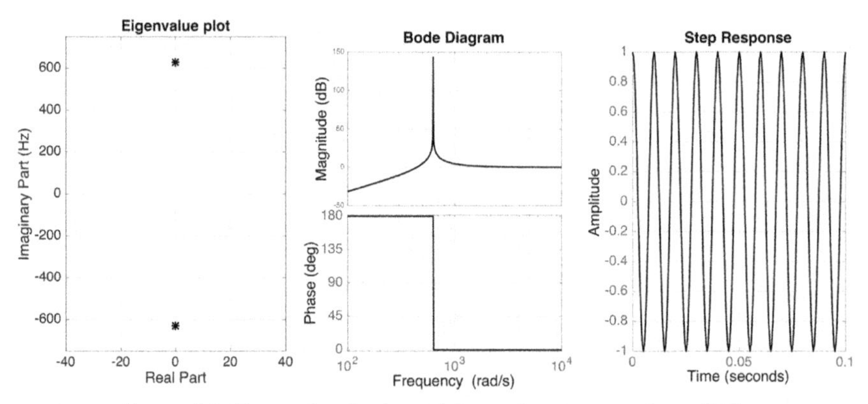

Figure 4.8 Eigenvalue, Bode and dynamic response plots of T8.

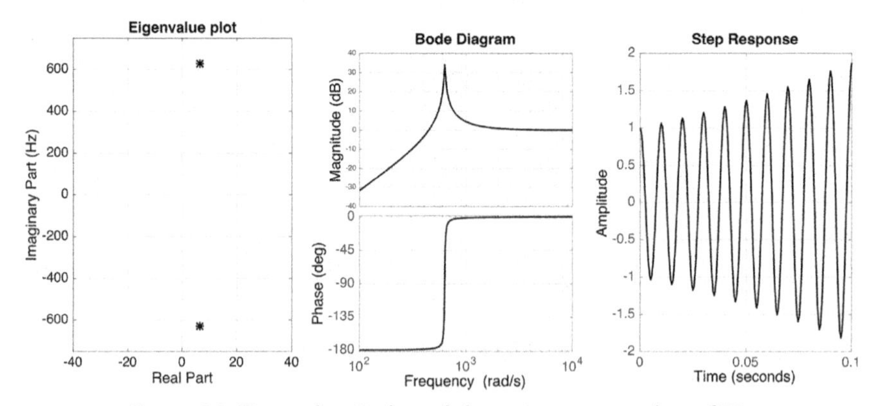

Figure 4.9 Eigenvalue, Bode and dynamic response plots of T9.

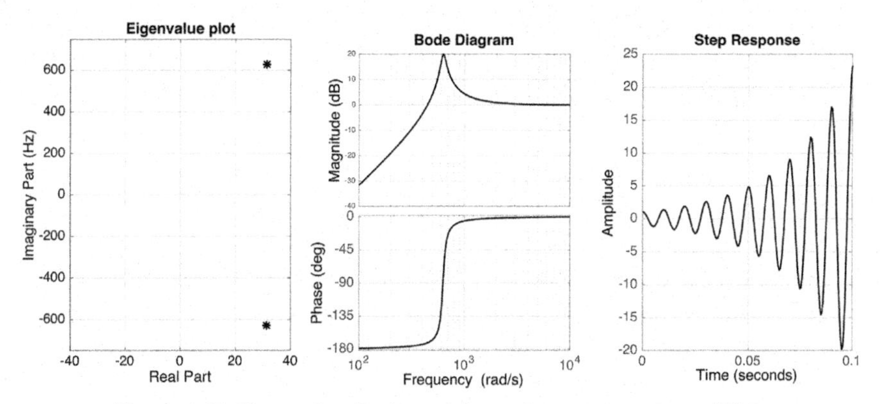

Figure 4.10 Eigenvalue, Bode and dynamic response plots of T10.

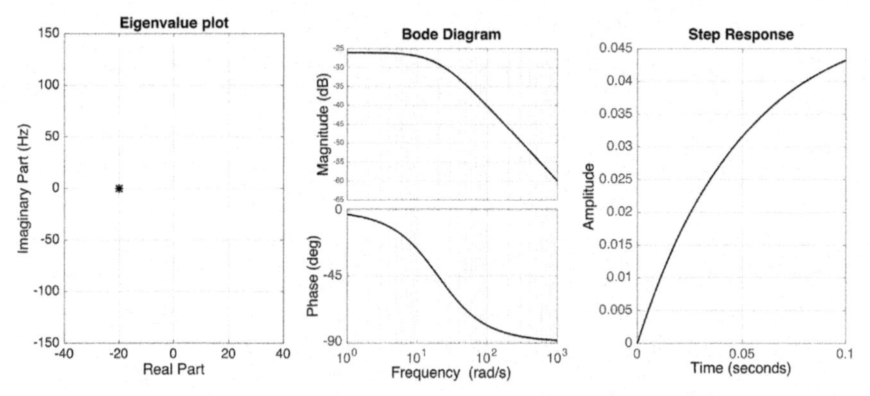

Figure 4.11 Eigenvalue, Bode and dynamic response plots of T11.

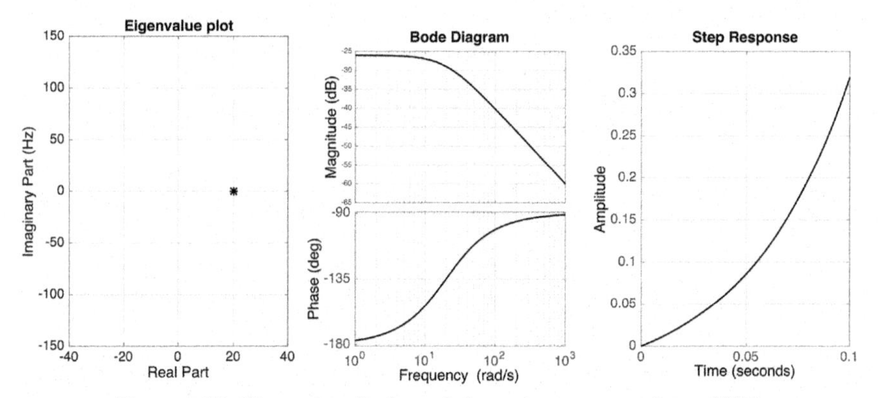

Figure 4.12 Eigenvalue, Bode and dynamic response plots of T12.

response of a linear system for a range of frequencies. Note that the frequency in eigenvalue plot and Bode plot is in rad/sec, which can be converted to Hz by dividing with 2π.

- Identify stable, marginally stable and unstable systems from the location of their eigenvalues and compare the dynamic response of the transfer functions.
- Select two transfer functions having same frequency and different damping ratios (for example, T1 and T2), and compare the location of eigenvalues and shape of the Bode plots. How does the behaviour at corner frequency (frequency of the eigenvalue) change with change in damping ratio? As the damping reduces, the gain plot is sharp with higher gain at the corner frequency, which is approximately equal to $1/2\zeta$ for a second order system. Similarly, the slope of phase plot is large for small damping ratio. Now compare the Bode plot with the step response.
- Now look at two transfer functions with different frequency and same damping ratio (for example, T2 and T7). Bode plot has same shape, but the peak occurs at a different frequency. But as the frequency increases, the real part of eigenvalue moves further away from the imaginary axis for a constant damping ratio. Now compare settling time for the two systems. Although the damping ratios are same, high frequency oscillation settles faster in $\frac{1}{|\sigma|}$ s. This is an important point to note. Damping ratio alone cannot provide a full picture about an oscillatory mode. From a stability perspective, a mode with low frequency should have higher damping ratio, as it will take more time to settle.
- Finally, note how an eigenvalue at the real axis provide an exponential response.

For further investigation, use *tf* command to build a transfer function, *eig* command to get eigenvalues, *bode* command to plot Bode plot and *step* to plot step response. For example, for T1, copy the following code to Matlab command window. (Sometimes Matlab may return an error if the command is copied and pasted to the command window. Try correcting '*' part of the command to rectify error.)

T1 = tf([1,0,0], [1, 6.291, 3958]), subplot(1,3,1), plot(eig(T1),''), subplot(1,3,2), bode (T1), subplot(1,3,3), step(T1)*

Another useful Matlab command to learn and work with at this point is *ginput,* which is used to get data points from Matlab figure. *[a,b] = ginput(n)* helps to acquire n coordinate points from a Matlab plot. In our case, if we like to measure frequency from a time domain response plot, use *ginput(2)* command to obtain x coordinates at two consecutive peaks. Find difference in x-axis points between two points and obtain its inverse, which is the frequency of the signal.

4.2 Power system model for analysis

For the remaining part of this chapter, we will use the SMIB Simulink model developed in Chapter 3. First, we will linearize the nonlinear Simulink model and obtain a linear state space representation. Using the state space model, eigenvalue and participation factor analysis will be explained. Later, the conversion from state space to transfer function and zeros-poles-constants representation will be discussed. We will discuss how to plot root locus, Bode, Nichols chart and Nyquist plots for the linear model of the synchronous machine. Finally, commands for controller design that improve the stability of the system are discussed.

Before we move forward, we need to make modifications in the synchronous machine model developed in Chapter 3. Save the Simulink model of the synchronous machine as smib_synch.slx. Run smib_run.m and then run smib_synch.slx. Is the simulation running without any error and producing the expected results? Now add an input port and an output port as shown in Fig. 4.13, which makes it a single-input and single-output (SISO) system. The input, Vref, changes the excitation reference voltage, whereas the output gives rotor slip.

Why is the combination of Vref and Slip selected? It is because we would like to design a controller to improve damping of the complex mode in the system. The mode is observable in Slip output and controllable through Vref input, meaning that this combination has high magnitude of residue corresponding to the complex mode. Such a controller is called power system stabilizer. We will discuss more about signal actuator selection in Chapter 10.

4.3 Linearization and state space representation

Small signal stability or stability under small disturbances is an important requirement for the operation of a power system. Linear techniques are widely used for the small signal stability analysis. So, the first task in our hand is to get a linear representation of the nonlinear power system model. Matlab provides easy commands to obtain a linear model as in Eq. (4.1) from the Simulink model.

The command [A,B,C,D] = *linmod('Simulink_File_Name')* performs a block-wise linearization of the Simulink model and gives the state space representation. Alternatively, use sys = *linmod('Simulink_File_Name')* to get a structure containing the state matrices, state names, input and output names and information about the operating point.

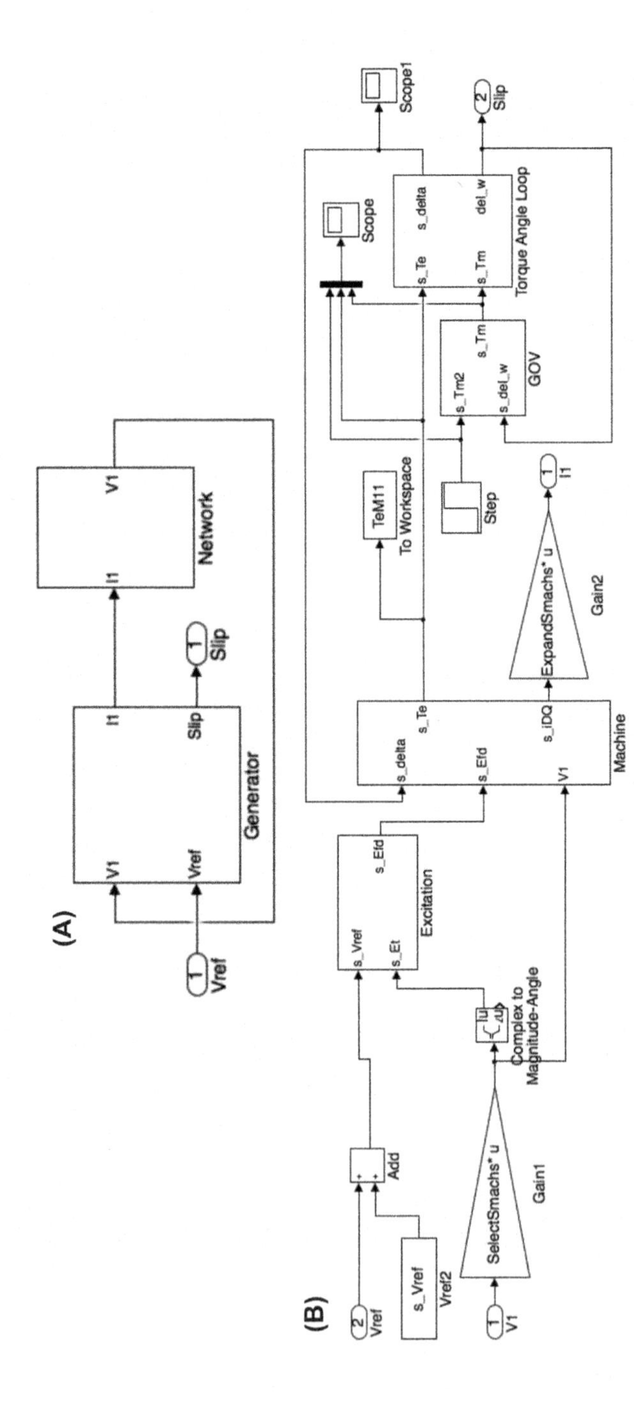

Figure 4.13 Modifications in the synchronous machine model developed in Chapter 3. (A) shows over all model and (B) shows inside view of generator block.

Table 4.2 State matrices of the smib_synch system.

A								B
0.000	−0.009	0.054	−0.055	−0.055	−0.126	0.009	0.000	0.000
0.000	−9.635	−6.479	−0.013	−0.013	−0.369	0.000	0.000	0.000
0.000	−18.403	−29.580	0.223	0.223	6.295	0.000	0.000	0.000
0.000	0.001	−0.005	−1.137	0.863	−0.124	0.000	25.000	0.000
0.000	0.030	−0.182	30.672	−35.994	−4.403	0.000	0.000	0.000
376.990	0.000	0.000	0.000	0.000	0.000	0.000	0.000	0.000
−900.000	0.000	0.000	0.000	0.000	0.000	−5.000	0.000	0.000
0.000	2.818	−16.910	−11.327	−11.327	4.877	0.000	−50.000	50.000
C								D
1.000	0.000	0.000	0.000	0.000	0.000	0.000	0.000	0.000

```
A1 = [ 0, -0.0090, 0.0542, -0.0554, -0.0554, -0.1264, 0.0085, 0;
 0, -9.6345, -6.4787, -0.0131, -0.0131, -0.3694, 0, 0;
 0, -18.4033, -29.5801, 0.2234, 0.2234, 6.2952, 0, 0;
 0, 0.0009, -0.0051, -1.1373, 0.8627, -0.1238, 0, 25.0000;
 0, 0.0304, -0.1824, 30.6722, -35.9945, -4.4029, 0, 0;
 376.9911, 0, 0, 0, 0, 0, 0, 0;
 -900.0000, 0, 0, 0, 0, 0, -5.0000, 0;
 0, 2.8183, -16.9097, -11.3267 -11.3267 4.8769 0 -50.0000];
B1 = [0, 0, 0, 0, 0, 0, 0, 50]';
C1 = [1, 0, 0, 0, 0, 0, 0, 0];
D1 = [0];
sys1 = ss(A1,B1,C1,D1);
```

Script 4.1 Matlab command to make system matrices (stable case).

Let us test the command. Run smib_run.m script. Type *[A,B,C,D] = linmod('smib_synch')* on the command window. Table 4.2 shows the output matrices obtained.

This is an SISO system, as we connected one input port and one output port in the model. The input name is Vref, and output name is Slip.

- Mathworks continuously improve their product and add new functionality or methods in the software. Often, there is more than one method to perform a calculation. For linearization, *linearize* command gives linear approximation of a Simulink model. You are encouraged to explore more on *mathworks.com/help/slcontrol/ug/linearize.html.*
- If you could not finish the chapter on synchronous machine modelling, but want to follow the remainder of this chapter, the system matrices can be formed using the Script 4.1.

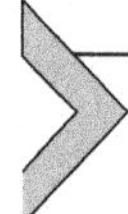

4.4 Eigenvalues, eigenvectors and participation factor

Eigenvalues appear often in power system stability analysis as they provide information on oscillatory behaviour and the stability of the system. Let us obtain the eigenvalue of the *smib_synch* system using the command *[PHI, L, PSI] = eig(A)*. The diagonal elements of L are the eigenvalues $(\lambda_1, \lambda_2, \ldots)$ of A, which are obtained by the command *lambda=diag(E)*. The eigenvalues of the *smib_synch* system are listed in Table 4.3.

The size of the A matrix is eight; hence, there are eight eigenvalues. All the eigenvalues have a negative real part, meaning the system is stable. Two of the eigenvalues (fifth and sixth) are a complex pair, which will demonstrate an oscillatory behaviour. The frequency of oscillation f and damping ratio ζ are given in Table 4.4.

The frequency and damping ratio of the oscillation are 1.14 Hz and 9.67%, respectively. Now if we go back to Fig. 3.29 in Chapter 3, it shows a response of the *smib_synch* system to verify that the oscillation has frequency 1.14Hz and the initial amplitude settles to 37% in approximately $\frac{1}{2\pi\zeta}$ cycles.

Table 4.3 Eigenvalues of the smib_synch system.

Index	Eigenvalues (*e*)
1	−53.80
2	−34.40
3	−16.8 + 9.6i
4	−16.8 − 9.6i
5	−0.70 + 7.2i
6	−0.70 − 7.21i
7	−2.95
8	−4.98

Table 4.4 Commands for Frequency and damping ratio calculation.

Equations with respect to fifth eigenvalue (λ_5)	Matlab command for fifth eigenvalue
$f_5 = \omega_5/2\pi$ (Hz)	Frequency = imag(lambda(5))/(2*pi)
$\zeta_5 = -100*\frac{\sigma_5}{\sqrt{\sigma_5^2+\omega_5^2}}$ (%)	Zeta = −100*real(lambda(5))/abs(lambda(5))

4.4.1 Exercise 4.2

This exercise is to understand the relationship between frequency and damping ratio, which will help to interpret stability in power system context. Look at Script 4.2, which is the state space representation of two systems. Determine the eigenvalues and damping ratios of the two systems. They have different frequencies, but the same damping ratio. Run the program in Script 4.2, which will plot the step response of the two systems. As damping ratio is same, both responses decay at same rate and settle to 37% of the initial amplitude in approximately $\frac{1}{2\pi\zeta}$ cycles, which is independent of the frequency. Hence, both plots look similar. However, notice the time axis to find that one system settles faster than the other. From a stability perspective, a mode with low frequency should have higher damping ratio, as it will take more time to settle.

With eigenvalues, the command *eig* also provides right eigenvectors (PHI) and left eigenvectors (PSI). The eigenvector matrices can be related to $\backslash phi$ and ψ as

$$\mathrm{PHI} = [\phi_1,\ \phi_2 \ldots \phi_n]$$

$$\mathrm{PSI} = \begin{bmatrix} \psi_1^T & \psi_2^T & \cdots & \psi_n^T \end{bmatrix}$$

Let us use them to find participation factors corresponding to the eigenvalues. The participation factor analysis is widely used in power system analysis, which helps to identify the state variables participating in an oscillatory behaviour or mode. The relative participation of *k*th state in the *i*th eigenvalue is given by $pf_{ki} = \left| \left| \mathrm{PHI}_{ki} \right| \left| \mathrm{PSI}_{ki} \right| \middle/ \sum_{k=1}^{k=n} \left| \mathrm{PHI}_{ki} \right| \left| \mathrm{PSI}_{ki} \right| \right|$. Script 4.3 is a function to calculate the participation factors for one eigenvalue. Copy the program and save as find_pf.m.

Now use the command pf5 = find_pf(PHI,PSI,5) to obtain the participation factors corresponding to the fifth eigenvalue. Table 4.5 shows the right eigenvector, left eigenvector and participation factors corresponding to the fifth eigenvalue (λ_5).

```
% Exercise system 1: State Space Representation
a1 = [-0.3945, -39.45; 1.0, 0]; b1 = [1;0]; c1 = [0, 0.3945]; d1=0;
% Plot step response of Exercise system 1
subplot(2,1,1), step(ss(a1,b1,c1,d1)), grid on
% Exercise system 2: State Space Representation
a2 = [-3.945, -3945; 1.0, 0]; b2 = [1;0]; c2 = [0, 3.945]; d2=0;
% Plot step response of Exercise system 1
subplot(2,1,2), step(ss(a2,b2,c2,d2)), grid on
```

Script 4.2 Exercise 4.2 program.

```
function [pf] = find_pf(PHI, PSI, index)
% vright and vleft are right and left eigenvectors obtained from eig
% command, and index is the index of eigenvalue for which
% Participation factor calculation is carried out.

% find the dot product of two eigenvectors
dotproduct = abs(PHI(:,index)).*abs(PSI(:,index));
% normalise with sum of the product
pf = dotproduct/sum(dotproduct);
```

Script 4.3 Function to calculate participation factor.

Table 4.5 Eigenvectors and participation factors.

Right eigenvector (PHI_5)	Left eigenvector (PSI_5^T)	Participation factor (pf_5)
−0.0042 − 0.0071i	0.9997 + 0.0000i	0.4433
−0.0209 − 0.0751i	0.0004−0.0101i	0.0423
−0.0350 + 0.1123i	0.0027 + 0.0046i	0.0338
−0.0658 + 0.0165i	−0.0079 − 0.0046i	0.0332
−0.0168 − 0.0148i	−0.0007 + 0.0007i	0.0012
−0.3472 + 0.2554i	−0.0006 − 0.0171i	0.3934
0.8882 + 0.0000i	0.0005 + 0.0009i	0.0483
−0.0071 − 0.0169i	−0.0036 − 0.0029i	0.0045

Before we discuss about the participation factor, use the command *sys = linmod('smib_synch')* to obtain the state names of the system. The command outputs a structure *sys*, and one of its elements is *StateName* as given below.

Now here is an important information. The value of *StateName* depends on the names of integrator representing states in the Simulink model. It is always important to use meaningful names for all parameters in a program. Table 4.6 shows system structure and state names obtained through linmod command.

Table 4.6 sys stucture of smib_synch.

sys		sys.StateName
a: [8x8 double]	1	'Smib_synch/Generator/Torque Angle Loop/s_wr '
b: [8x1 double]	2	'Smib_synch/Generator/Machine/ Machine Equations/s_Ed_dash/s_Edd'
c: [1 0 0 0 0 0 0 0]	3	'Smib_synch/Generator/Machine/ Machine Equations/s_Psi2q/s_Psi2_q'
d: 0	4	'Smib_synch/Generator/Machine/ Machine Equations/s_Eq_dash/s_Eqd'
StateName: {8x1 cell}	5	'Smib_synch/Generator/Machine/ Machine Equations/s_Psi1d/s_Psi1_d'
OutputName: {'smib_synch/Slip}	6	'Smib_synch/Generator/Torque Angle Loop/s_delta'
InputName: {'smib_synch/Vref'}	7	'Smib_synch/Generator/GOV/s_Tm1'
OperPoint: [1x1 struct]	8	'Smib_synch/Generator/Excitation/ s_Efd//s_Ka'
Ts: 0		

Table 4.7 Commands for transfer function, ZPK and residue.

Representation	Matlab command
$H_1(s) = \frac{b_0 s^m + b_1 s^{m-1} + \ldots b_m}{a_0 s^n + a_1 s^{n-1} + \ldots a_n}$	[num,den] = ss2tf(A,B,C,D)
$H_2(s) = K\frac{(s-z_1)(s-z_2)\ldots(s-z_m)}{(s-p_1)(s-p_2)\ldots(s-p_n)}$	[Z,P,K] = ss2zp(A,B,C,D)
$H_3(s) = \frac{r_1}{s-p_1} + \frac{r_2}{s-p_2}\ldots\ldots + k$	[r,p,k] = residue(num,den)

Now compare the state names against the participation factor. Which state has got the highest participation, and which has got the lowest? In the case of e_5, the highest participation factor is 0.4434, which is first in the list. The first state is s_wr, which is the generator speed (ω) state. The second highest is 0.3934, which is sixth in the list. The sixth state name is s_delta, which corresponds to the generator rotor angle (δ) state. This shows that the mode has high participation from the generator speed and generator rotor angle state; hence, this mode is referred as electromechanical mode. A signal with high participation factor is preferred to design feedback controller. In this case, the speed signal has higher participation factor for e_5, and it can be used as feedback signal for a controller to improve damping of e_5. The controller design is explained in Section 4.5. In Chapter 10, using a four-machine, two-area test system, we will explore the importance of electromechanical modes in power system stability.

4.5 Transfer function and ZPK representation

So far, we have used the state space representation of the linearized system model. As mentioned in this Section 4.1, there are other ways of representing the linear system such as the transfer function model and zero-pole-gain model. The state space representation we have can be converted to a transfer function, ZPK and partial fraction expansion form using the commands given in Table 4.7.

It is possible to convert one form to another very easily, and you are encouraged to find more commands from the Matlab manual.

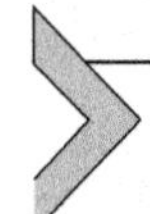

4.6 Root locus, Bode plot, Nichols plot and Nyquist plot

In this section, we will demonstrate the use of some of the frequency domain tools in control systems and the Matlab commands to plot them.

```
A2=[0,0.0006,-0.0036,-0.0364,-0.0364,-0.0387,0.0085,0;
0,-9.5962,-6.7087,-0.0005,-0.0005,0.0216,0,0;
0,-19.0566,-25.6603,0.0078,0.0078,-0.3679,0,0;
0,0.0000,-0.0002,-1.1067,0.8933,-0.0883,0,25.0000;
0,0.0011,-0.0064,31.7610,-34.9057,-3.1398,0,0;
376.9911,0,0,0,0,0,0,0;
-900.0000,0,0,0,0,0,-5.0000,0;
0,4.1140,-24.6843,-12.5545,-12.5545,8.3370,0,-50.0000];

B2=[0,0,0,0,0,0,0,50]';
C2=[1,0,0,0,0,0,0,0];
D2=[0];
sys2=ss(A2,B2,C2,D2);
```

Script 4.4 Script for making state matrices for Case 2 (unstable case).

To get	Command
Bode plot	`bode(A,B,C,D) or margin(A,B,C,D)`
Magnitude and phase information	`[Mag, Phase] = bode(A,B,C,D)`
Root Locus plot	`rlocus(A,B,C,D)`
Nichols chart	`nichols(A,B,C,D); ngrid;`
Nyquist plot	`nyquist(A,B,C,D);`

Script 4.5 Commands for frequency domain plots of the system.

To better appreciate these plots, we will define two cases, a stable case and an unstable case of the *synch* system. The state matrices of the stable system (Case_1) are given in Script 4.1. This is obtained from smib_synch system by keeping all the parameters the same as discussed in Chapter 3. The system structure is obtained using the command sys1= linmod('smib_synch');.

To make the Case_2, change the line matrix to line = [1 2 0.00 0.09 0.0000 1.0 0.0]. This makes the resistance of the line zero and series reactance 0.09 pu. The system structure is obtained using the command sys2= linmod('smib_synch');. The state matrices of the unstable case are given in Script 4.4. The eigenvalues of the two systems are plotted in Fig. 4.14. Note that the sys1 is a stable system, whereas the sys2 is an unstable system with two complex eigenvalues with positive real parts.

Note: If you could not complete Chapter 3, Script 4.1 has the Case 1 state matrices and Script 4.4 has the Case 2 state matrices.

Using these two systems, we will discuss Bode plot, root locus, Nichols plot and Nyquist plot and extract important stability information. The Matlab commands to obtain these plots are given in Script 4.5.

Before we move ahead, let us define some of the important terms. Fig. 4.15 shows a simple feedback system. $G(S)$ is the plant transfer function, $H(s)$ is controller transfer function and K is a controller gain. In our case, $G(s)$ can be the sys1 or sys2 system.

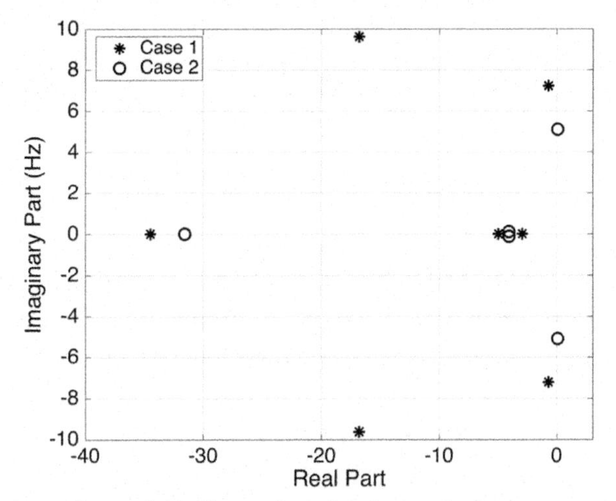

Figure 4.14 Eigenvalue of smib_synch systems.

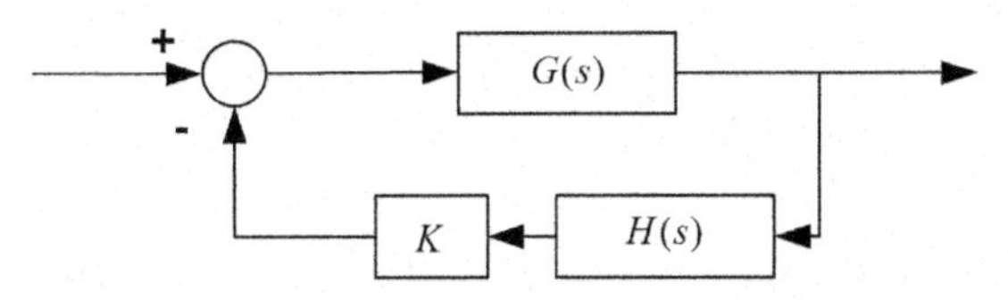

Figure 4.15 Closed loop system block diagram.

For ease of explanation, assume $G = G(s)$ and $H = KH(s)$. For controller design perspective, we can define the open loop transfer function as $G_{ol} = GH$, which can be represented as $G_{ol} = \frac{(s-z_1)(s-z_2)\ldots(s-z_m)}{(s-p_1)(s-p_2)\ldots.(s-p_n)}$, where p and z are poles and zeros. If none of the poles have positive real part, the open loop system is stable. What about the closed loop system stability? The closed loop transfer function is $G_{cl} = \frac{G}{1+GH}$, whose poles must not have positive real part.

Now let us think about root locus plot. It essentially sketches trajectory of poles of closed loop transfer function G_{cl} on a complex plane, as K changes from zero to infinity. There will be as many lines as number of poles. The lines terminate at a zero or at infinity. So, if there are n poles and m zeros, $n - m$ lines will terminate at infinity. How helpful is such a plot? We can easily identify instability by checking any lines moving towards the right half of the complex plane. If it is moving towards unstable region, the plot can show for what range of values the system is stable or unstable. Is not it useful? We will demonstrate this little later. One important thing to mention here is that the K parameter in root locus analysis need not be the controller constant, but it could be any other parameters of the plant transfer function. Root locus plot can show the effect of uncertainty of any selected parameter on the stability of the closed loop system.

Apart from the effect of the changes in controller gain or system parameters, engineers are often interested in the frequency response of the system. By frequency response we mean the response of the system to an input frequency. This can be defined in terms of gain of the transfer function $|G|$, which corresponds to the change in amplitude between input and output and phase response $\angle G$, which corresponds to phase shift between input and output waveform.

Bode plots are exactly the same. It has two plots, one for gain and another for phase. However, there is a slight modification required. The magnitude is plotted in decibels, frequency is in logarithmic scale and phase angle is plotted in degrees. However, the Matlab plot can be changed to linear scale if required. A Bode plot can tell us about the closed loop stability of a system. But for that, we need to define few important terms.

Gain crossover frequency (ω_g) – Frequency at which gain plot crosses 0 dB.

Phase crossover frequency (ω_p) – Frequency at which phase plot crosses 180 degrees.

Gain margin – Change in gain of open loop system that can make closed loop system unstable, which is negative of gain at ω_p.

Phase margin – Change in phase of open loop system that can make closed loop system unstable, which is 180+phase at ω_g.

If both the gain and phase margins are positive, the system will be stable when we close the loop. Higher the margins, the greater the system can withstand changes in system parameters.

Nichols chart is a similar tool like Bode, which plots magnitude in dB against phase angle in degrees. We will compare this plot in the next section.

Now we come to last of the four plots we are discussing in this section. The Nyquist is another similar but very useful tool in control system, especially when the open loop system contains positive poles, meaning it is unstable. So, what is a Nyquist plot and what is Nyquist stability criteria?

A closed loop system G_{cl} is stable when all its eigenvalues have positive real part. Incidentally, poles of G_{cl} are same as zeros of $1 + GH$. This means for stability, number of zeros of $1 + GH$ in the right half of the complex plane must be zero. The Nyquist diagram plots $real(G(jw))$ against $imaginary(G(jw))$ as the frequency changes from infinity to zero to infinity.

Before interpreting the plot, we need to obtain number of poles of GH with positive real part. Let us assume it is P. Let Z be number of zeros of $1+GH$ with positive real part. For a stable system, Z will be zero. Now count the number of times (N) Nyquist plot encircles $-1+j0$. An anticlockwise encirclement is counted at one and clockwise is counted as -1. Number of zeros of $1+GH$ with positive real part or number of unstable poles of G_{cl}, $Z = P + N$. If Z is positive, the system is unstable.

4.7 Analysis of stable system

Copy the Script 4.1 to Matlab to build sys1 representing the stable case of synch_smib. Let us assume controller transfer function $H(s) = 1$, meaning feedback gain is K. The root locus, Bode, Nichols and Nyquist plots can be obtained by commands rlocus(sys1), bode(sys1), nichols (sys1) and nyquist(sys1), respectively. Fig. 4.16 shows the four plots for reference.

4.7.1 Root locus plots

Let us closely examine the root locus plot. All the eigenvalues are at negative part of the complex plane suggesting a stable system when $K=0$. As K

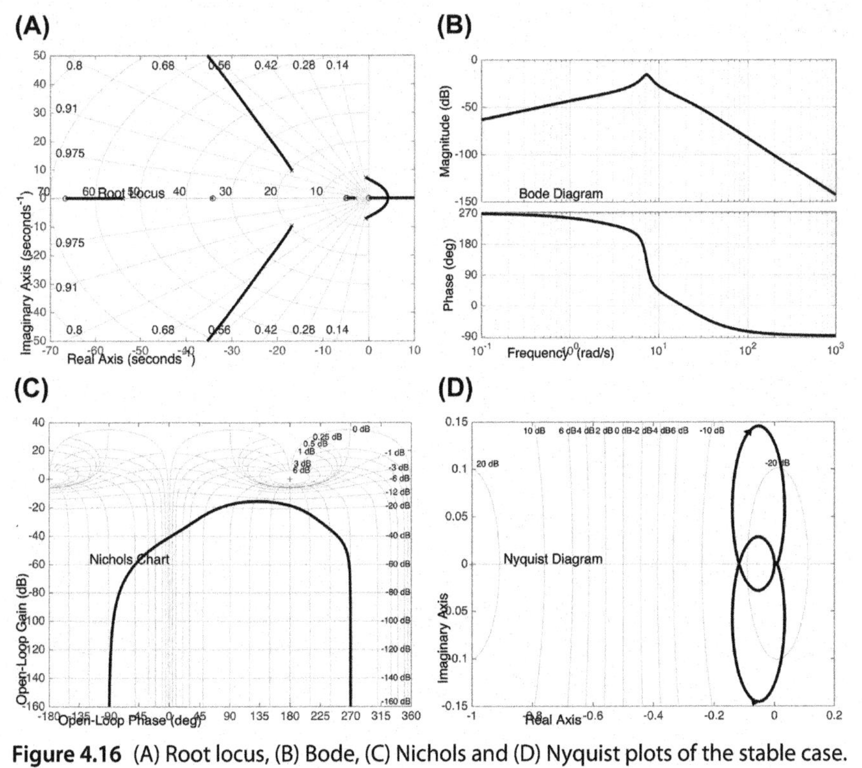

Figure 4.16 (A) Root locus, (B) Bode, (C) Nichols and (D) Nyquist plots of the stable case.

increases, two of the modes shift towards positive half of the complex plane causing instability. Find values of K for which the system is stable by clicking on the plot when it touches the imaginary axis, which would be 8.31. Although the system is closed loop stable, the damping ratio is reduced as the feedback gain increased.

4.7.2 Bode, Nichols and Nyquist plots

Generate the plots using the command discussed above. Right click on the plot, then select characteristics → *all stability margins*. Dots appear on the plots at frequency equal to gain crossover frequency and phase crossover frequency. Click on the dots to display gain margin, phase margin and stability information. The gain margin in this case is 18.6 dB. How is this value connected to gain $K = 8.31$ obtained from the root locus plot (20log(8.31)= 18.4, which is not exactly equal but close to 18.6 dB)? What is the connection between the phase crossover frequency and the frequency at which root locus plot crosses imaginary axis? Compare all the three plots and look for similarities. Zoom the Nyquist plot near $-1+j0$ point and notice that the plot is not encircling the -1 point. As we know already, there are no unstable poles for the open loop system; the number of unstable zeros of $1+GH$ is zero. This indicates that the closed loop transfer function has no poles at the right half of the complex plane. Hence, closed loop system is stable.

4.8 Analysis of unstable system

Now let us make the four plots for sys2 in Script 4.4. First of all, assume $H(s) = 1$. The plots for the unstable system are shown in Fig. 4.17. The root locus plot shows two eigenvalues on the positive side of the complex plane. The feedback is not stabilizing the system, as the root locus plots are moving further away from the imaginary axis. All the other plots indicate that the system is unstable. In the Nyquist plot, the graph does not encircle -1, but open loop system has two poles at the right-hand side, meaning there are $2 + 0 = 2$ unstable zeros in $1 + GH$ or two unstable poles in the closed loop system. So in this case, the closed loop system is unstable.

Consider a very simple transfer function $H(s) = \frac{25}{2s^2+3s}$. Use commands *num =25, den = [2,3,0]* and *H=tf(num,den)* to build the transfer function. Now plot root locus, Bode, Nichols and Nyquist plots for the system sys2 with a feedback controller $H(s)$. Use following commands to obtain the plots.

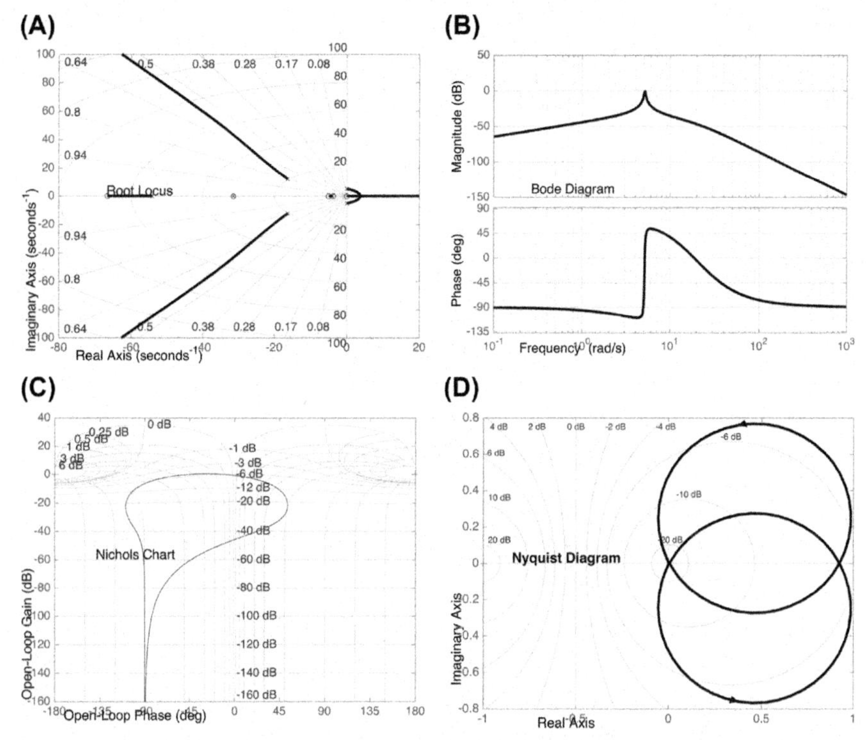

Figure 4.17 (A) Root locus, (B) Bode, (C) Nichols and (D) Nyquist plots of the unstable case.

```
% build plant and controller transfer function
Gs = ss(A2,B2,C2,D2); Hs = ss(tf(num,den));
rlocus(Gs*Hs)% root locus plot
bode(Gs*Hs) % bode plot
Nichols(Gs*Hs) % Nichols plot
Nyquist(Gs*Hs) % Nyquist plot
```

Root locus plot shows the poles moving from unstable region to stable region. Other plots indicate that the system is stable. There are two open loop poles in the unstable region and two encirclements of -1 point by the graph, indicating no unstable poles in the closed loop system.

Before we move forward, use the *residue* command introduced in Section 4.3 and find the angle of residue corresponding to the complex mode. How is the angle related to the angle of departure from the complex pole? The angles corresponding to the complex poles $(0.0671 \pm 5.1025i)$ for sys2 are $\pm 151.5^0$. The controller transfer function $H(s)$ has phase angle −160.3 degree at the complex pole. See the connection between these numbers. $H(s)$ gives sufficient rotation for the root locus to move towards left in closed loop instead of towards right as in open loop.

4.8.1 Linear system analyzer

The above plots can be put together by using Linear System Analyzer app. It is an interactive user interface for analyzing the time and frequency responses of linear systems and comparing such systems. Before you start, make a linear system representation using the command sys = ss(A,B,C,D), now type the command *linearSystemAnalyzer* on the Matlab command window, which opens a user interface. On the file menu of the Linear System Analyzer app, click import to import a sys system. Use the Edit→configuration menu to generate various plots of the system. The resulting menu is shown in Fig. 4.18.

4.9 System response

In Section 4.1, we have seen the step response of few transfer functions. Obtaining system response to different input signals is important for control system design. Table 4.8 lists few of the useful Matlab commands to plot the system response of the system.

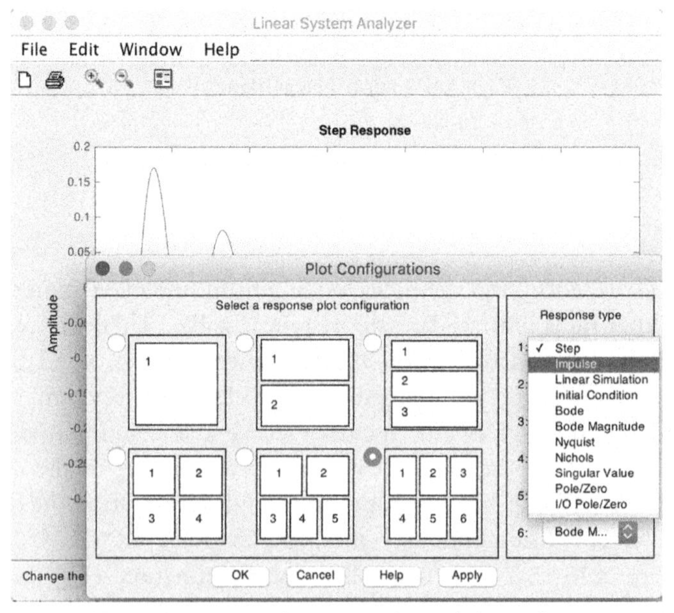

Figure 4.18 Linear System Analyzer dialogue box.

Table 4.8 Matlab commands for system response.

To get	Command
Step response	sys = ss(A,B,C,D); step(sys)
Impulse response	sys = ss(A,B,C,D); impulse(sys)
Response to a defined input	lsim(ss(A,B,C,D))

4.10 Controller design

The objective of this section is not to discuss the theory behind controller design, but to discuss some of the Matlab commands for controller design and demonstrate their working using the *smib_synch* system model. An important thing to remember always is that these commands could generate controller transfer function that can provide a stable closed loop response in Matlab. However, their application in real life physical system depends on many factors. For example, in a power system, the selected feedback signal may not be local to the controller, feedback signal may not be easily measurable, it may be subjected to delay or have noise or the controller transfer function simply cannot be realized. However, we will assume that we are in a perfect world, free from these limitations, and focus on the application of the commands using the system available.

Before we start, let make sure we have the [A2, B2, C2, D2] matrices for the Case 2 of *smib_synch* system ready. Get transfer function representation of system G using the command *[num,den]=ss2tf(A2, B2, C2, D2); G=tf(num,den);*.

4.10.1 PI controller

We will try to develop a simple PI controller for the system. First define a PI controller in parallel form $\left(K_p + \frac{K_i}{s}\right)$ using the command cntr = *pid(1,1);*.

Similarly, a standard PI controller of form $(K_p\left(1 + \frac{1}{sT_i}\right))$ can be defined using the command *pidstd*. Now tune a controller for the plant G using the command K = pidtune(G,cntr);. For the system in Case 2, the parameters obtained are $K_p = -21$, $K_i = -1.4$. However, it also returns a warning, 'Warning: Cannot keep loop gain above 1 at low frequency and below 1

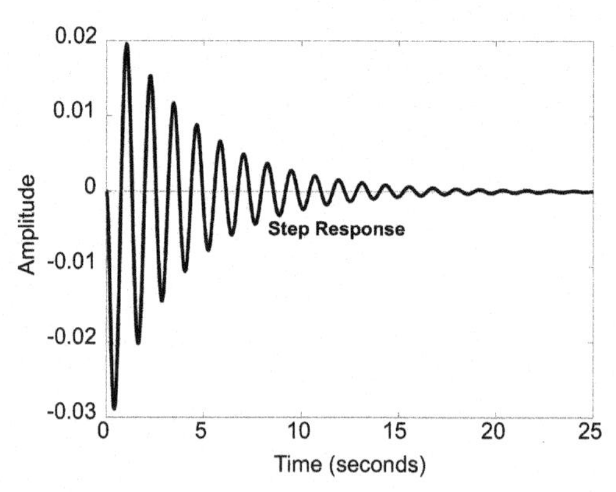

Figure 4.19 Step response of the system using a PI controller.

at high frequency. Try adding integrators to the loop'. What it means is that the Matlab could not find parameters for a PI controller that can stabilize the system. We can tune the controller using Matlab's graphical PI controller tool called *pidTuner*. Try it using the command *pidTuner(G,K)*. Play with different options to get required performance. Tuned controller parameters are shown at the bottom of the graphical window. The PI controller parameters obtained for this system using this approach are not the best, but the exercise is intended to demonstrate the tools.

For further work, let us use controller transfer function used in Section 4.8. Use commands *num =25, den = [2,3,0]* and *K=tf(num,den)* to get controller transfer function. Now test the controller performance using a step response. Use the feedback command to obtain the closed loop system transfer function and use the step command to obtain the step response.

step(G) % step response of the open loop system
hold on
CLG = feedback(G,K); % to get closed loop transfer function
step (CLG) % step response of the closed loop system

The response of the system is shown in Fig. 4.19.

The step response of the closed loop system is stable compared with the open loop system. Now plug the controller into the Simulink model as shown in Fig. 4.20. A gain block equal to −1 is used to create a negative feedback, as the sum block is inside the generator block, where *Vref* input is connected has two positive inputs. Now run simulation and observer response. The response of the system to step change in turbine mechanical input is shown in Fig. 4.21.

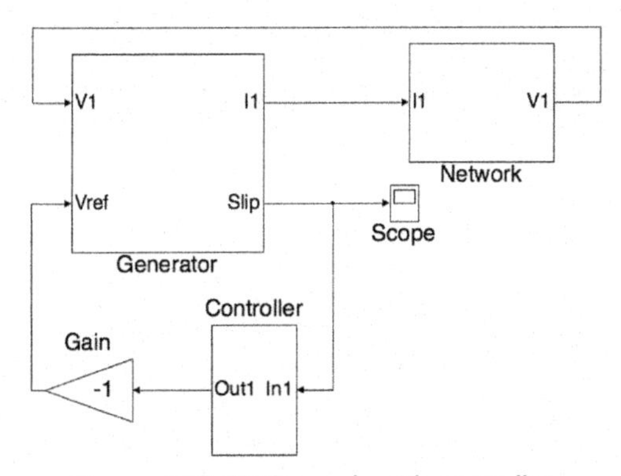

Figure 4.20 SMIB_synch with controller.

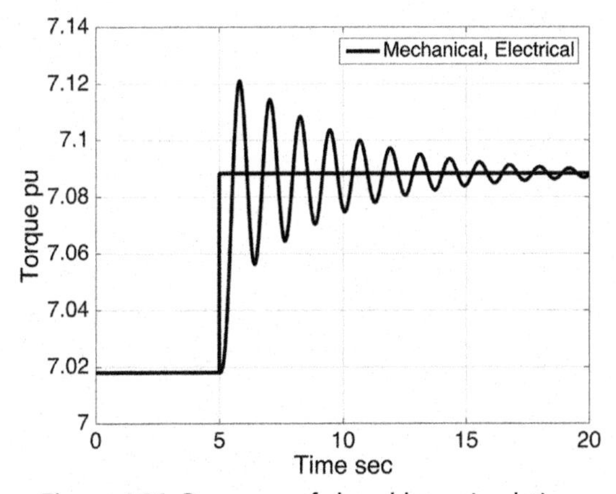

Figure 4.21 Response of closed loop simulation.

4.10.2 Control System Designer

This app helps to graphically tune an SISO control loop directly from Simulink. We will demonstrate the functionality using simple steps using the closed loop simulation model in Fig. 4.20.

Step 1: Define input and output signal points as shown in Fig. 4.22. Right click at the output of step block and select Linear Analysis Point→Input Perturbation. Similarly, right click on the line connecting Torque Angle Loop block and Speed output port and select Linear Analysis Point→Output Measurement.

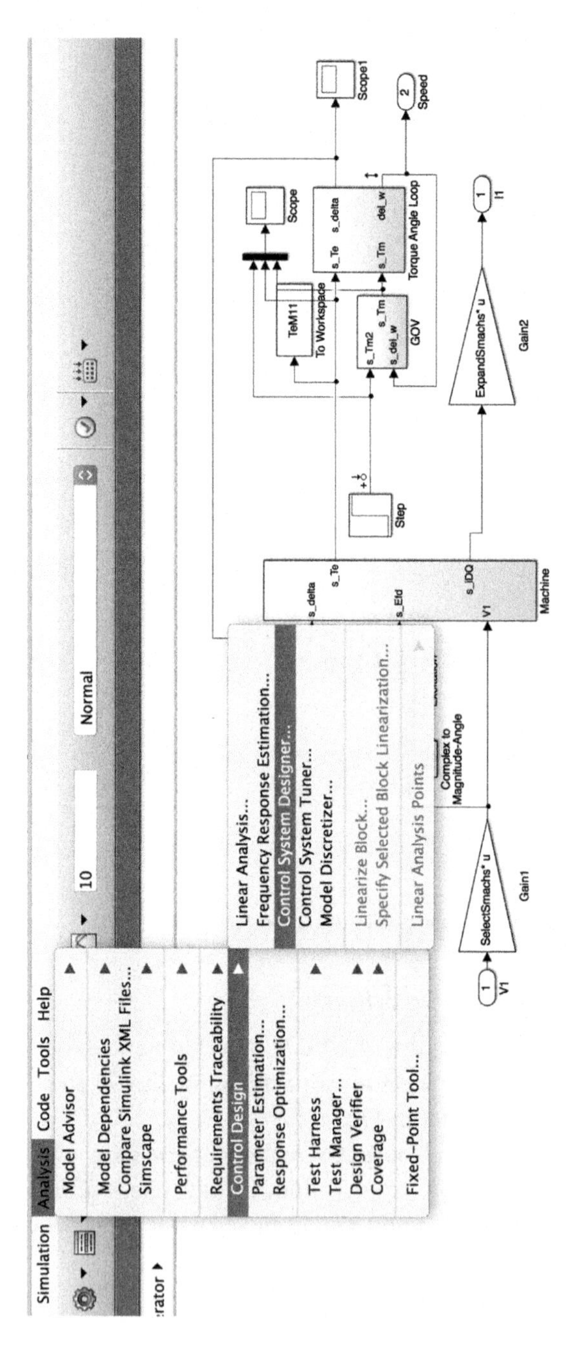

Figure 4.22 Simulink block showing input and output signal points.

Step 2: Select Analysis→Control Design→Control System Designer. This will open Control and Estimation Design Manager as shown in Fig. 4.23. There are many options to choose in this app. We will make few simple choices to demonstrate the concept.

Step 3: Click Select Blocks button that opens a new dialogue. Select the blocks to tune as shown in Fig. 4.23.

Step 4: Select closed loop signals as shown in Fig. 4.24. Click Tune Blocks. A Design Configuration Wizard opens. Click Next. We will not make any selection here. Click Next once again. The window will look like Fig. 4.25. Make selections as shown in the figure and click Finish button.

We have selected the Pole/Zero option, which will open a window as shown in Fig. 4.26. Make sure the real time update check box is selected. There will be another open window with Compensator Editor Option as shown in Fig. 4.27. Now move the slider right and left, which changes the controller parameter, and notice the changes in the complex pole. When gain is more than −4, the pole moves to right-hand side of the imaginary axis. Now click the Update Simulink Block Parameter button to update the gain parameter in the Simulink.

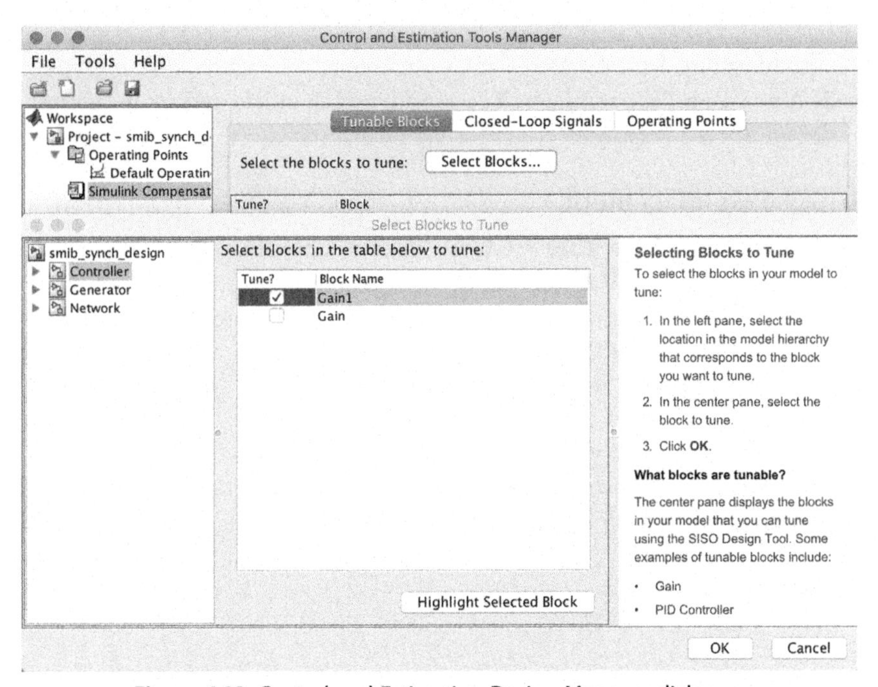

Figure 4.23 Control and Estimation Design Manager dialogue.

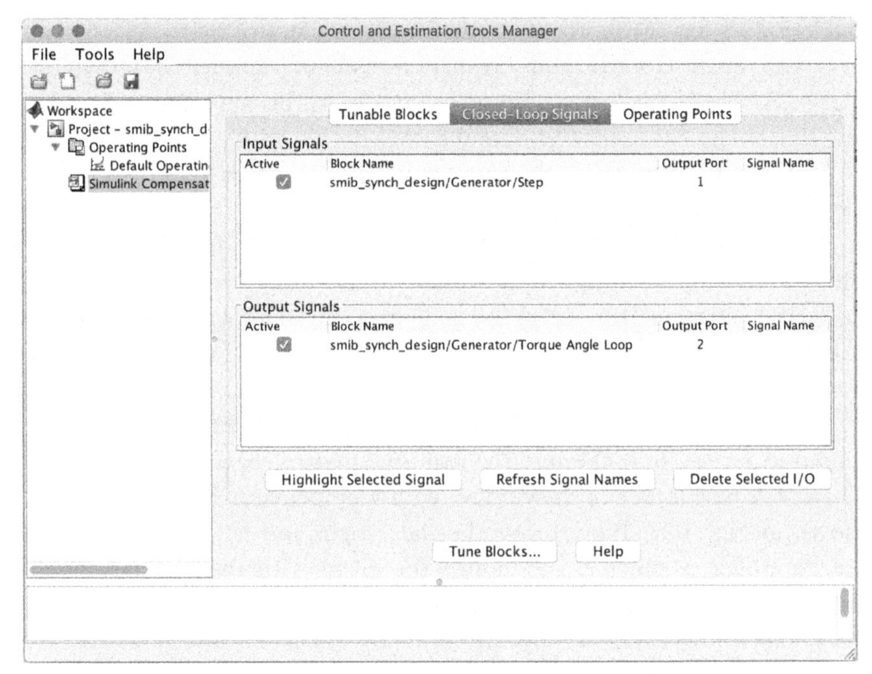

Figure 4.24 Control and Estimation Design Manager closed loop signals.

We have just demonstrated the Control System Designer app to graphically tune controller parameter. You are advised to utilize options in the app to design desired response.

4.10.3 Pole placement

Pole placement uses a state feedback control approach. Suppose in the system ($\dot{x} = Ax + Bu$) all the states are available for feedback, command *place* can calculate a controller K that can place the poles into desired location p. The resulting closed loop system is of the form ($\dot{x} = (A - BK)x + Bu$).

```
K = place(A,B,p)
```

Here p is the target location of the closed loop poles.

In Case 2, two of the eigenvalues have positive real part. Let us try to find a controller, which can place the eigenvalues in the left half side of the imaginary axis. Define p as

```
p=[-20.0, -33.3333, -30.1276, -20.6264, -0.1 + 5.1025i, - 0.1 -
5.1025i, -4.5838, -5.5078]';
```

p is selected such that all elements in p have negative real part.

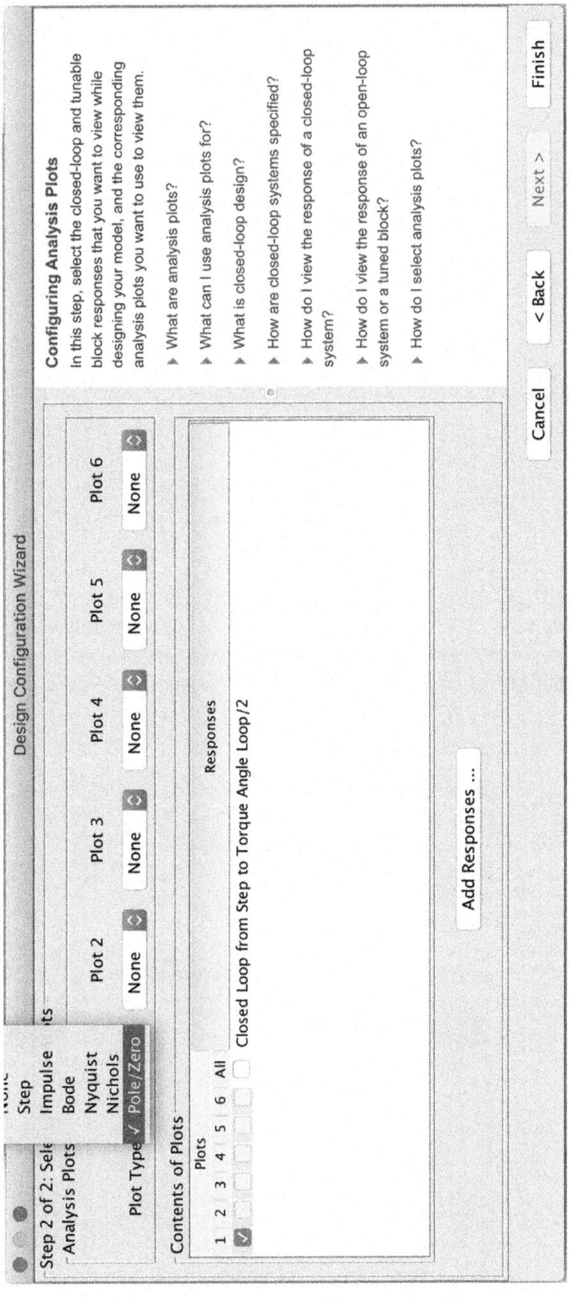

Design Configuration Wizard
Step 2 of 2: Sele
Analysis Plots
None
Step
Impulse
Bode
Nyquist
Nichols
✓ Pole/Zero
Plot Type
Plot 2 Plot 3 Plot 4 Plot 5 Plot 6
None None None None None
Contents of Plots
Plots
1 2 3 4 5 6 All
Responses
Closed Loop from Step to Torque Angle Loop/2
Add Responses ...
Configuring Analysis Plots
In this step, select the closed-loop and tunable block responses that you want to view while designing your model, and the corresponding analysis plots you want to use to view them.
▸ What are analysis plots?
▸ What can I use analysis plots for?
▸ What is closed-loop design?
▸ How are closed-loop systems specified?
▸ How do I view the response of a closed-loop system?
▸ How do I view the response of an open-loop system or a tuned block?
▸ How do I select analysis plots?
Cancel < Back Next > Finish

Figure 4.25 Control and Estimation Design Manager: Analysis plots selection.

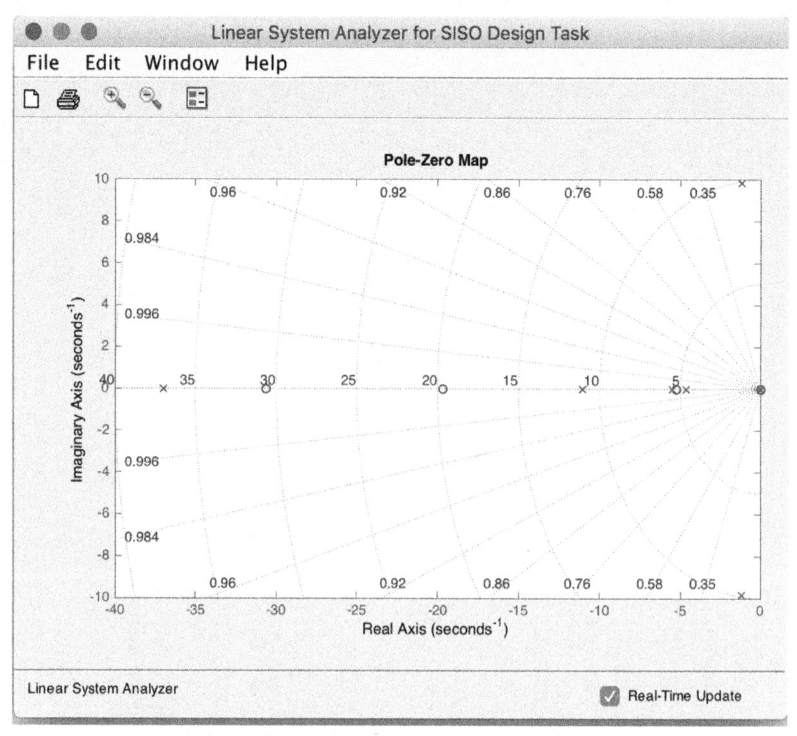

Figure 4.26 Control and Estimation Design Manager: Pole/Zero plots.

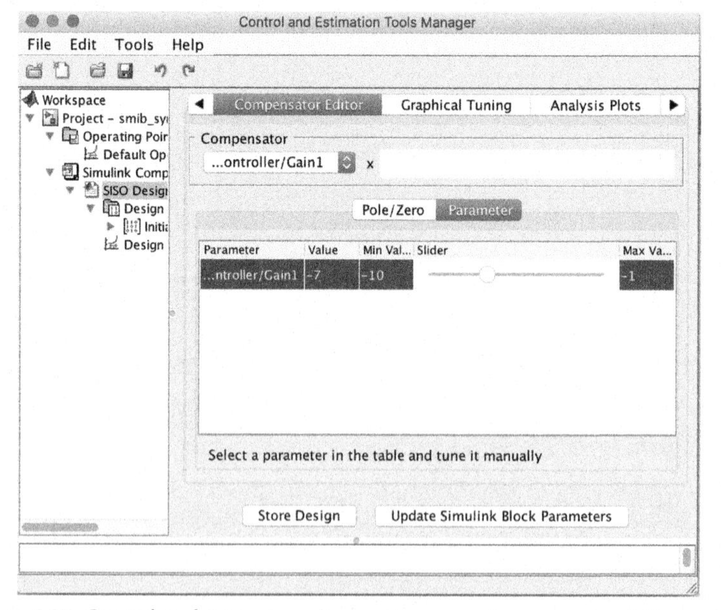

Figure 4.27 Control and Estimation Design Manager: Compensator Editor option.

```
% Obtain system matrices
[A2,B2,C2,D2]= linmod('smib_synch');    % or copy Script 4.4
% Define required location of closed loop poles. In this selection,
% only the two complex poles are different from the open loop poles.

p=[ -54.0351, -31.5198, -16.3485+12.1154i, -16.3485-12.1154i,...
-0.1+5.1025i, -0.1-5.1025i, -4.0757+0.1129i, -4.0757-0.1129i];

% Find controller gain using pole placement
K = place(A2,B2,p);

% Plot eigenvalue of open loop and closed loop system
plot(eig(A2),'*'), hold on, plot(eig(A2-B2*K),'*')
```

Script 4.6 Program for controller design and validation using pole placement.

Find the controller using $K=$ place($A2$, $B2$, p).

Using the controller, the closed loop system representation becomes, $\dot{x} = (A - BK)x$. The eigenvalues of the closed loop system can be obtained using *eig(A-B*K)*. The process can be summarized in the Script 4.6.

4.10.4 Linear Quadratic Regulator controller

Linear Quadratic Regulator is another well-known state feedback controller design technique. We will demonstrate its use here.

For a system $\dot{x} = Ax + Bu$ and control law $u = -Kx$, the optimum gain matrix K is obtained by minimizing the quadratic cost function.

$$J(u) = \int_0^{\infty} \left(x^T Qx + u^T Ru + 2x^T Nu\right) dt$$

where Q and R are cost functions. For simple case, specify $R=1$ and $Q = C'*C$, which sets equal weightage to input and states. The commands for the controller design are given in the Script 4.7.

```
% Obtain system matrices
[A2,B2,C2,D2]= linmod('smib_synch');    % or copy Script 4.4
% Find Q and R matrices
Q = C2'*C2;
R = 1;
% Find controller gain using LQR
K = lqr(A2,B2,Q,R);

% Plot eigenvalue of open loop and closed loop system
plot(eig(A2),'*'), hold on, plot(eig(A2-B2*K),'*')
```

Script 4.7 Controller design program using Linear Quadratic Regulator.

4.11 Conclusions

This chapter describes several control system tools in Matlab, which can be used for stability analysis and controller design for power system application. The SMIB test system model is used to demonstrate the commands. It should be kept in mind that these commands are just tools to help simplify complex mathematical calculations, and physical constraints of the system must be taken into consideration while using these tools.

CHAPTER FIVE

Load modelling

Power system loads have different characteristics. The large single loads are industrial in nature, such as steal and aluminium smelting plants. In the commercial sector, loads for supermarkets or railway transportation are large by nature. Residential loads are always treated in the small load category. Smaller embedded generations, such as roof top solar panels and small-scale wind turbines installed alongside households, are included in the loads as a negative contribution for steady-state analysis purposes, although this representation is a crude simplification. With more and more power electronic interfaces to the loads, the profile and characteristic of loads is evolving. The overall characteristic of loads seen by the network is further changing with the increasing penetration of electric vehicles. The size of the load and the load mix changes continuously. One of the tasks involved in power system modelling is to appropriately capture the behaviour of this mixture of loads while limiting the complexity of the modelling. This is achieved using a simplified representation of the loads in the network, where the cumulative behaviour of a segment of the low voltage part of the grid including its loads is represented by a single load at the grid supply point.

This chapter discusses various types of load models, with a detailed description of the frequently used ZIP (constant impedance/current/power) modelling. A four-machine system is used to investigate the effect of various load characteristics on the small signal behaviour of the system poles. While carrying out a dynamic simulation in Matlab/Simulink, ZIP load modelling introduces an algebraic loop in the system, and this chapter discusses two techniques for breaking the loop and compares both techniques for validation purposes.

5.1 Types of loads

The modelling details of the loads depend on the focus of the study and the available data as well as power voltage characteristics. We distinguish between dynamic load and ZIP load representation. In dynamic loads, motoring loads are represented by their dynamic equations, analogous to the modelling approach taken for generators. Usually frequency and motor

Simulation of Power System with Renewables
ISBN: 978-0-12-811187-1
https://doi.org/10.1016/B978-0-12-811187-1.00005-6
© 2020 Elsevier Inc.
All rights reserved.

slip are state variables. The modelling details for motoring loads are readily available in (Kundur, 1994). The dynamic representation of motoring loads and more detailed representation of loads in general may be required for long-term stability, voltage stability and interarea oscillation studies or systems dominated by motoring loads. For many studies, the response of the load clusters, as seen from the point of aggregation, to changes in voltage is very fast, allowing for the simplified model representation as ZIP loads. This is particularly true for loads that are mainly nonmotoring in nature.

For the ZIP load model, the load is represented by an algebraic relationship, related to a change in load bus voltage. Z stands for constant impedance, I for constant current and P for constant power. In Chapter 3, the load was represented by a pure constant impedance load, which allows the load to be integrated with the network impedance matrix. Heating loads are typically constant impedance in nature. In ZIP load representation, the real and reactive parts of the load can be independent of Z, I or P type or a composite mixture of all three types. If the real or reactive part of a load of mixed ZIP characteristic is to be modelled, this can be done using an exponential model or a polynomial model (Kundur, 1994). The ZIP load representation can further be extended to include a dependency of the load behaviour on load bus frequency.

The ZIP parameters can be determined from measurements and can be presented in form of tables, and efforts have been recently made to update ZIP parameters near real time as the mix of loads connected to the network evolves (Zhao et al., 2018) and to use the ZIP parameters at substation level to identify the types of loads connected to the system (Asres et al., 2019). A more detailed knowledge of the load dynamics is also attained in power systems through the widespread introduction of smart metres, and more flexibility in load behaviour will be achieved through demand side response services. When sufficient data are unavailable, a frequent assumption is that the real power is a constant current load and the reactive power a constant impedance load (Kundur, 1994).

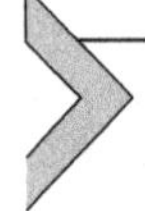

5.2 Descriptions, key equations and integration of ZIP model

As mentioned earlier, the real and reactive parts of loads are modelled independently to allow for different voltage-dependent behaviours of the real and reactive loads. The constant impedance, constant current and constant power (Z, I and P) loads are modelled by setting the variable a in (5.1) for real and b in (5.2) for reactive power to 2, 1 or 0, respectively, where $|V|$ is the magnitude of the load bus voltage and $|V_0|$, P_0 and Q_0 are the initial conditions (ICs) of the bus voltage magnitude and power. A simple and commonly used assumption models the real part of the load as constant current and the reactive power as constant impedance. What are values of the variables a and b in this case? Can you write down the equations for the real part of the load power P_L and reactive part of the load power Q_L using this assumption?

$$P_L = P_0 \left(\frac{|V|}{|V_0|} \right)^a \tag{5.1}$$

$$Q_L = Q_0 \left(\frac{|V|}{|V_0|} \right)^b \tag{5.2}$$

For composite loads, where both P_L and Q_L can be of a mixed ZIP characteristic, the values of a and b can be chosen as nonintegers to match the voltage characteristic of the curves for the loads P and Q values with a change in voltage. This is called the exponential model.

Alternatively, mixed ZIP characteristics of loads can be described by a polynomial model, which is simply derived from the equations above. In the polynomial model, P_L and Q_L consist of three loads each, a constant impedance, a constant current and a constant power one. The three loads are weighted out of 100% by the proportion of the load they constitute, where the proportions are denoted by p_{1-3} and q_{1-3}. Please have a go at simplifying the expression and also write down the polynomial expression for Q_L.

$$P_L = P_0 p_1 \left(\frac{|V|}{|V_0|} \right)^2 + P_0 p_2 \left(\frac{|V|}{|V_0|} \right)^1 + P_0 p_3 \left(\frac{|V|}{|V_0|} \right)^0 \tag{5.3}$$

Now we have an expression for the power of the loads as shown in (5.1), (5.2) or (5.3) and the expression you just derived for Q_L. We need to write down an expression for the load current, which is readily derived as (5.5) using the equation for complex power S_L in (5.4). In general, you will find in power system simulations that the network takes currents as inputs and gives bus voltages as outputs, while all other network components take the bus voltage as input and give current injections as outputs. The generator convention is used in the model developed in this book, and as such, the power consumed by loads is negative.

$$S_L = VI_L^* \tag{5.4}$$

$$((P_L + jQ_L)/V)^* = I_L \tag{5.5}$$

In the case of implicit load modelling, which can be used only for constant impedance type loads, the impedance Z is representative of the network impedance Znet and impedance type loads Zload, such that Z equals Znet plus Zload.

If you refer to Script 2.5, as it was used in Chapter 2, this is how the impedance type loads were implicitly represented as part of the impedance matrix. As we wish to model the loads explicitly, we only use the last line and do not include the loads in the Y-matrix, when we build the impedance matrix.

For the load model, the initial load current will be required, which can be readily determined from the load bus voltages and complex load powers using (5.5) as shown in Script 5.2.

The generator modelling has been first introduced in Chapter 3, and as shown in Fig. 5.1 through the solid box and line, the generator model is

```
%The loads are assumed to be impedance type loads here and are included in
%the Y matrix. The Y matrix is then inverted to find the impedance matrix.

        YPL = PL./V.^2;
        YQL = QL./V.^2;
        Y = Y + diag(YPL-j*YQL);
        Z = inv(Y);
```

Script 5.1 Implicit representation of Z-load model in impedance matrix.

```
%Calculates the complex voltage value for all buses.

        vol = bus_sol(:,2).*exp(1i*bus_sol(:,3)*pi/180);

%PL and QL are the real and reactive power drawn by loads at all system
%buses.

        PL = bus_sol(:,6);
        QL = bus_sol(:,7);

% Load current

Iload= conj((PL +1i*QL )./vol);
```

Script 5.2 Initialization of parameters for explicit load modelling.

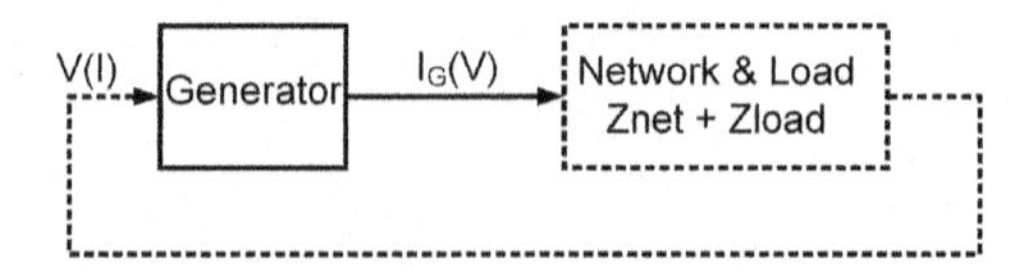

Figure 5.1 Implicit load model without algebraic loop.

made up of dynamic equations, whereas, as indicated by dashed lines, the network and load models are frequently represented by algebraic equations.

The main challenge of integrating these very simple ZIP load equations stems from algebraic loops. In time domain simulations, the initial starting point needs to be known for all variables. In dynamic equations, these are set via the ICs. An algebraic loop occurs, when an algebraic equation has a variable A as input and a variable B as output, and a second algebraic equation has variable B as input and variable A as output. As mentioned in Chapter 3, we have already used a simple load representation, where both the real and imaginary parts of the load were represented by impedance type loads. The implementation of this scenario is particularly simple, as the network impedance matrix and load impedance are merged, hence avoiding this algebraic loop. This is not the case for constant current and constant power loads in a network represented by an impedance matrix, making the implementation in our Simulink model more complex. This difference is visible contrasting Figs. 5.1 and 5.2. Fig. 5.1 shows that there exists no algebraic loop, when the loads are integrated as impedance loads in the Z matrix of the network. Fig. 5.2 shows the algebraic loop that is created when nonconstant impedance type (ZIP) loads are included in the simulation, and two subsystems with only algebraic equations become dependent on each others' outputs.

There are various ways to break the algebraic loop. The first one available in Simulink is to use the inbuilt IC block. This block can be inserted in the algebraic loop to tell Simulink what the initial value for the variable is. The second option is to introduce a very fast time constant in the loop. There is a third option, which is nontrivial to integrate into our Simulink models, which would involve creating a function block in the Simulink

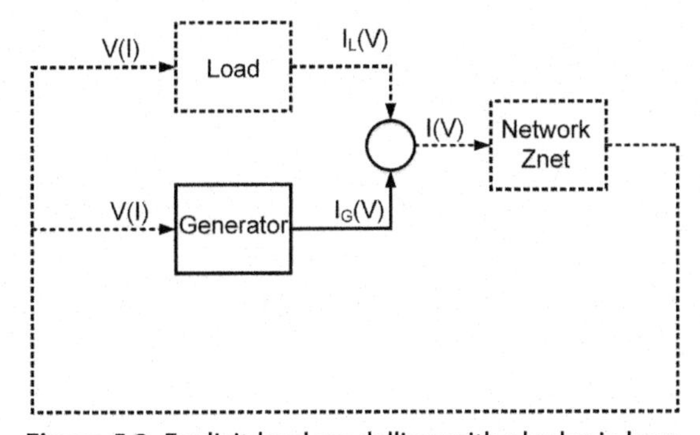

Figure 5.2 Explicit load modelling with algebraic loop.

program, which at each time step evaluates and determines the solution of the set of algebraic equations, before using this solution to evaluate the section with the differential equations, which is required for the next time step. Apart from the challenge of implementing such an inbuilt function and the requirement that all algebraic equations would need to be integrated in the inbuilt function block in the Simulink environment, the major downside is that linearization and analysis of the poles, in the way it was introduced in Chapter 4 using linmod, would not be possible with this inbuilt function implementation.

As such, we are left to choose between the IC block implementation and the delay function implementation. The IC block implementation has been shown to lead to erroneous eigenvalue results, when used directly with *linmod*. We will need to use a trick to achieve correct eigenvalue results using this method. The delay block method is the simplest method in terms of implementation and achieving reliable linearization results, as the algebraic loop is actually broken in this method. The short delay introduced for the purpose of breaking the loop may not reflect any actual delay in the system, and it does give rise to additional eigenvalues/poles, which are to be neglected in the stability analysis.

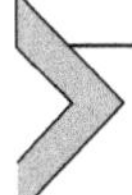

5.3 Study case: four-machine system using different load models

We will now build and run the simulation using both the IC block implementation and the delay block implementation. We can then compare the eigenvalue and time response results to establish their correct performance.

Common to both implementation methods is the load model part, so we will develop this first. Most of the script below should look familiar compared with the scripts used in Chapters 3 and 4. Compare them slowly, so you can highlight the changes made to accommodate the explicit load representation. Please pay particular attention to how this script does not include loads in the Y matrix.

With the amended Matlab initialization in place as described in Script 5.1 and 5.2, we can now build the Simulink model, including loads. Fig. 5.3 shows an example of explicit load representation. Looking at Fig. 5.3, with the gain parameter K equal to

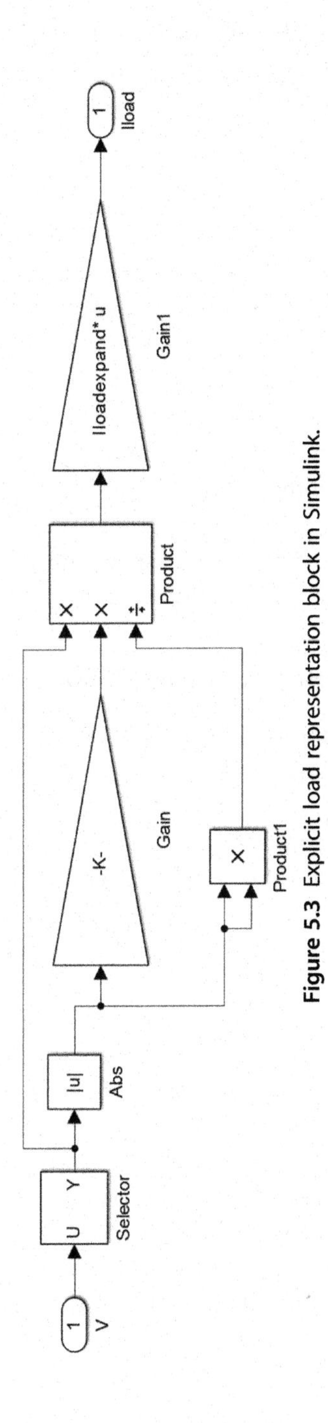

Figure 5.3 Explicit load representation block in Simulink.

$$\frac{[\mathbf{PL(7);PL(9)\text{-}i[QL(7);QL(9)]}]}{(\mathbf{abs[vol(7);vol(9)]})^2}$$

write down the formula and determine whether this model is for a constant current, constant power or constant impedance type load. Build the explicit load representation shown in Fig. 5.3 and include it in the four-machine system model developed in Chapter 3. For this simulation use two_area_synch.m and two_area_synch_model.slx, this time with the fault removed from the simulation and the governors connected. Please refer to Script 5.3 to ensure your initialization is suitable to follow this chapter. Save the Simulink model with the name *'Load_fourmach_withcontrol_sys_base'* adding the explicit load representation as a subsystem called loads, as shown in Fig. 5.4.

Let us now try to run the simulation and see what happens. An overview of the power system model (in our case a four-machine system) is shown in Fig. 5.4. Referring to Fig. 5.2, we can see that this system has an algebraic loop problem. When we run the Matlab script and Simulink model, Simulink accordingly throws up algebraic loop errors.

5.4 Initial condition block implementation

First let us explore the IC block implementation. Save the Simulink model *'Load_fourmach_withcontrol_sys_base'* with a new name *'IC_block_fourmach_withcontrol_sys_base'*. We can break the algebraic loop, by defining a starting point for the simulation, as shown in Fig. 5.5 using the IC block found in the library browser. As this block only handles real numbers, we split the current into real and imaginary parts, in the IC subsystem, as shown in Fig. 5.6. To receive reliable results from the linearization of the model we just made, we need to follow a few steps. Ensure the Simulink model is open and then run Script 5.4. Instead of running the model to compile, use the step forward bottom twice to move the model to the first nonzero time step. We ensure this first time step is close to time zero by setting the maximum time step to 1e-5. This setting is found under Simulation, Model Configuration Parameters. We can then run the two lines of code below. The first line defines the initial state vector, and the second line linearizes the model around the first nonzero time step. After this, eig(sys.a) will show the eigenvalues of the system under study.

When running simulations with IC block for altered systems, the order of states can be found by running:

sys = linmod('IC_block_fourmach_withcontrol_sys_base')

```
clear all

Include all four machine system data from Chapter 3 here.

% Obtain load flow solution
[Y] = form_Ymatrix(bus,line);
% calculate power flow solution
[bus_sol, line_flow] = power_flow(Y,bus, line);

% Synchronous machine initialisation

Smachs=[1;2;3;4];% buses with synchronous machines

% program to initialise the synchronous machines

Synch_parameter_sys_base_fourmach
generic_Sync_Init

%Calculates the vector of apparent power S injected by the
%generators for each system bus.
PQ = bus_sol(:,4)+1i*bus_sol(:,5);
%Calculates the complex voltage value for all buses.
vol = bus_sol(:,2).*exp(1i*bus_sol(:,3)*pi/180);
%Calculates the complex value of the current injected by the
%generators at each system bus
Icalc = conj(PQ./vol);
%PL and QL are the real and reactive power drawn by loads at
%all
%system buses.
PL = bus_sol(:,6);
QL = bus_sol(:,7);
%V is the voltage magnitude at all system buses.
V = bus_sol(:,2);

%Network impedance.
Z = inv(Y);

% Implicit matrix when loads are assumed to be impedance type
%loads here and are included in the Y matrix.

YPL = PL./V.^2;
YQL = QL./V.^2;
Ytotal = Y + diag(YPL-j*YQL);
Ztotal = inv(Ytotal);

%Voltage magnitude.

Volabs=abs(vol);

% Expand Y matrix.

Y = full(Y);

%Load powers.

PLload=PL;
QLload=QL;

%Time constant for transfer function implementation.

Tconst = 0.0001;

% Load current

Iload= conj((PL +1i*QL )./vol);

% Expanding the load current from buses with loads to all
%buses
Iloadexpand=zeros(11,2);
Iloadexpand(7,1)=1;
Iloadexpand(9,2)=1;

%Settings required for IC block implementation, a Simulink
%model needs to be open to run these lines of code

set_param(bdroot,'AnalyticLinearization','on')
set_param(bdroot,'BufferReuse','off')
set_param(bdroot,'RTWInlineParameters','on')
```

Script 5.3 Four-machine system script for explicit load representation.

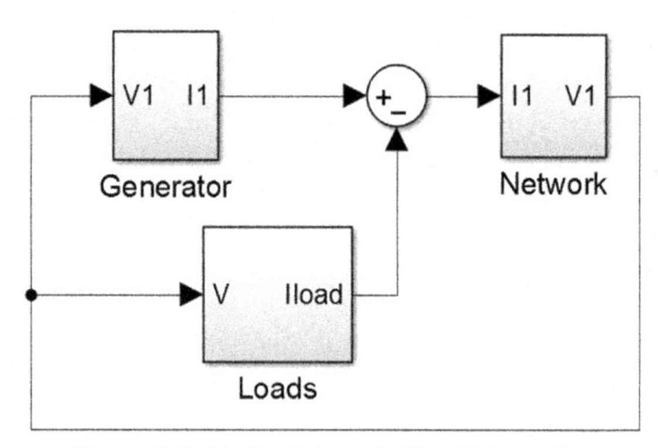

Figure 5.4 Algebraic loop in Simulink model.

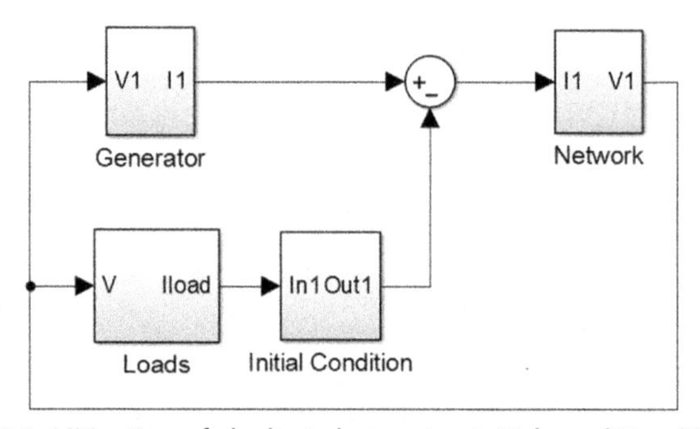

Figure 5.5 Mitigation of algebraic loop using initial condition (IC) block.

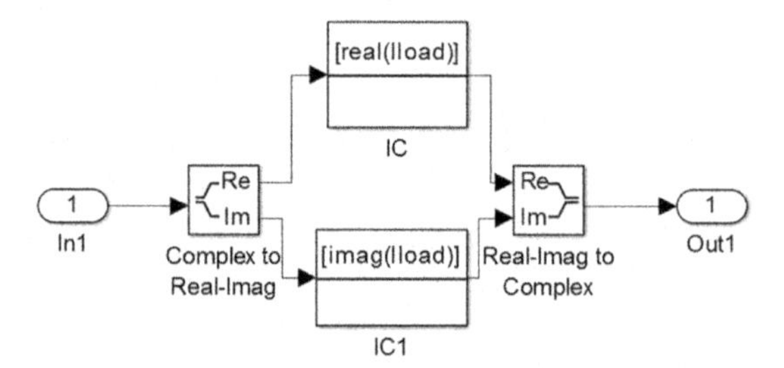

Figure 5.6 Initial condition (IC) block subsystem.

```
xstates= [s_Eqd; s_delta; s_Psi1d; s_Psi2q; s_Edd; s_Efd./s_Ka;...
s_ws;s_ws;s_ws;s_ws; s_Tm];
sys = linmod('IC_block_fourmach_withcontrol_sys_base',xstates,[],...
[1e-50.001 0])
```

Script 5.4 Linearization for initial condition (IC) block method.

The structure sys in the workspace lists the states in the order they are used in the linearization in sys.StateName. You can then use this order of states to create the vector states, required for the code above.

For the delay block implementation, start from the Simulink model `'Load_fourmach_withcontrol_sys_base'`, which has the algebraic loop, and save it with a new name `'delay_block_fourmach_withcontrol_sys_base'`. Include an additional subsystem, named delay, as shown in Fig. 5.7.

Fig. 5.8 shows the required components inside of the delay block. You can build your subsystem accordingly.

In the delay block, the integrators have integration constants of the real and imaginary parts of the load current Iload. Tconst is set to a very short time, here 0.001 s. For analysis purposes, dynamics with very fast time constants can be thought of as algebraic relationships, whereas dynamics with very slow dynamics can be thought of as constants; hence, introducing a very fast dynamic equation to break the algebraic loop is a suitable tool.

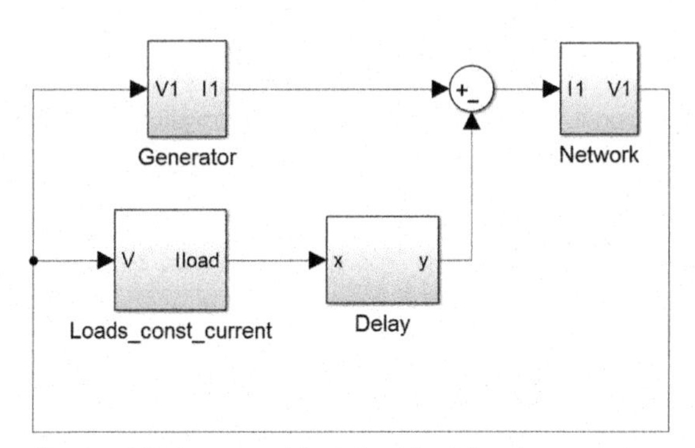

Figure 5.7 System with delay block implementation.

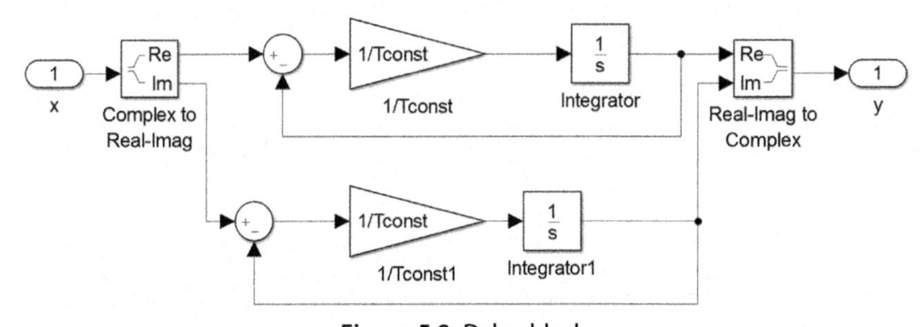

Figure 5.8 Delay block.

5.5 Comparison of results

Using both approaches, the IC block and the delay block, we can obtain the eigenvalues and plot these on one graph, using Script 5.5, to compare the results as shown in Fig. 5.9.

We can see the delay block implementation has introduced additional poles, with no physical meaning, which are on the far side of the left-

```
xstates= [s_Eqd; s_delta; s_Psi1d; s_Psi2q; s_Edd; s_Efd./s_Ka;...
    s_ws; s_ws;s_ws;s_ws; s_Tm];
sys = linmod('IC_block_fourmach_withcontrol_sys_base',xstates,[],...
    [1e-5 0.001 0])
% code first described in Script 5.4
eig(sys.a)
Atest=eig(sys.a);
% Atest stores eigenvalues of IC block implementation
scatter(real(eig(sys.a)), imag(eig(sys.a)),'ro');
hold on
sys = linmod('delay_block_fourmach_withcontrol_sys_base')
eig(sys.a)
Btest=eig(sys.a);
% Btest stores eigenvalues of delay block implementation
scatter(real(eig(sys.a)), imag(eig(sys.a)),'k*')
grid on
```

Script 5.5 Comparison of eigenvalues using initial condition (IC) block and delay method.

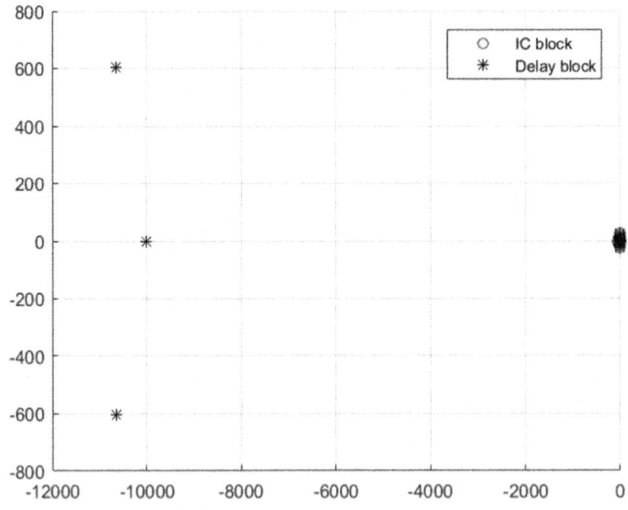

Figure 5.9 Poles of model with delay block (*black stars*) and initial condition (IC) block (*red circles*) (*black circle* in print version) method.

hand plain and should be neglected. These poles are easily identified. As we zoom into the poles we are actually interested in, shown in Fig. 5.10, we can generally see a good agreement between the poles of both the delay block and the IC block implementation. Applying two different methods of breaking, the algebraic loop for load modelling is a nice approach in providing a higher level of confidence in the accuracy of our modelling result.

We can take this a step further and calculate the mismatch between both sets of results, which is a measure of the accuracy of results. We can arrange the order of the eigenvalues, so we can compare corresponding poles, in this particular case we reorder according to Script 5.6. Please note that Script 5.6 is likely not to work for you, as it is specific to the order of eigenvalues. Use visual inspection of Atestsort and Btestsort, to reorder them accordingly.

This allows us to take the difference of the relevant matching poles and plot the size of difference in eigenvalues result between the IC block method and the delay block method. As can be seen in Fig. 5.11, maximum errors for the imaginary part are very small compared with the magnitude of the poles at a scale of 10^{-3}, and the maximum difference in the real part is less than 0.04, which is still a reasonable agreement viewed against the magnitude of the poles. The accuracy of the result for each pole can be determined from the discrepancy between the results of the two methods.

So far, we have demonstrated that constant current type loads can be modelled either using a delay block implementation or an IC block

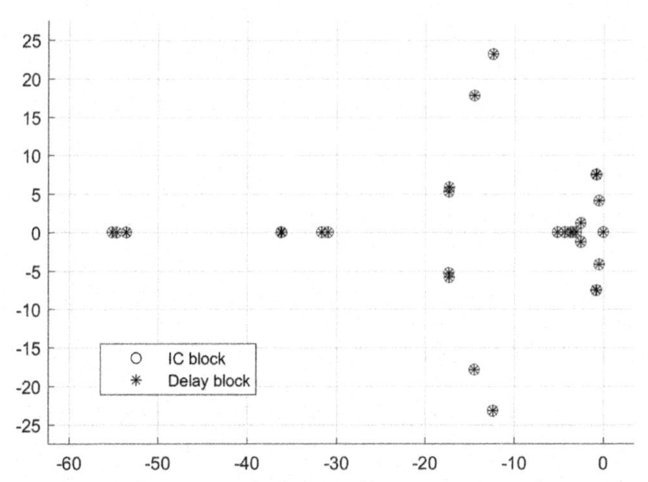

Figure 5.10 Zoomed view — Poles of model with delay block (*black stars*) and IC block (*red circles*) (*black circle* in print version) method.

```
Ctest=zeros(54,2);
%Ctest to compare Atest and Btest
%side by side
Btestsort= cplxpair(Btest);
%complex conjugate pairs grouped
Atestsort= cplxpair(Atest);
%complex conjugate pairs grouped
Ctest(1:54,1)=Btestsort;
%sorted Btest is first coloumn
Ctest(3:22,2)= cplxpair(Atestsort(1:20));
Ctest(43:54,2)= cplxpair(Atestsort(21:32));
%Atest is second coloumn, used visual inspection
%to align to be similar to Btest
%each reader needs to do this depending on their
%system
Ctest(:,3)=Ctest(:,1)-Ctest(:,2);
scatter(real(Ctest(:,3)), imag(Ctest(:,3)));
%scatter difference in Atest and Btest modes
%zoom to focus on actual mode difference
grid on;
```

Script 5.6 Reordering of poles for comparison of difference between initial condition (IC) block and delay block implementation.

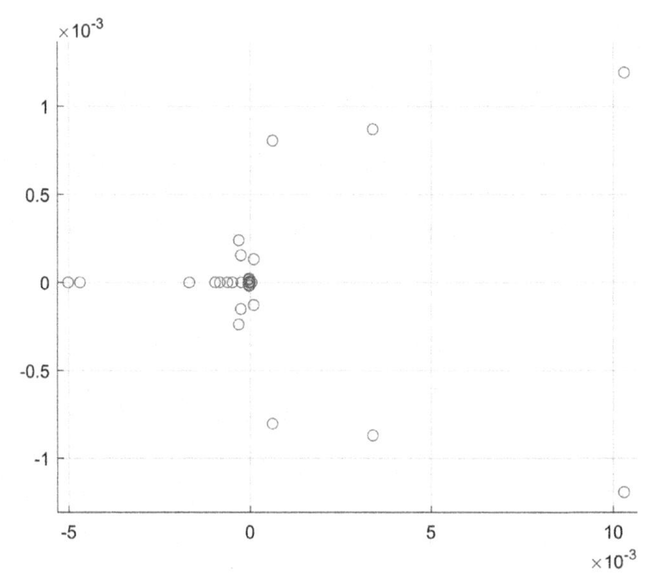

Figure 5.11 Difference of poles with delay block and initial condition (IC) block method.

implementation. Now we also extend this to the other two types of loads. We can further validate our approach to breaking the algebraic loop in this section, by comparing the eigenvalues of the constant impedance type loads for the implicit method and the explicit methods (IC and delay block). Finally, we can compare the three load types via their eigenvalues.

Figs. 5.12 and 5.13 show the explicit load representation for constant power and constant impedance type loads, respectively. The load model in Fig. 5.3 was for constant current load type, and as before, we make use of a mathematical identity, shown in (5.6).

$$\frac{1}{V^*} = \frac{V}{|V|^2} \tag{5.6}$$

Try to derive the subsystems in Figs. 5.12 and 5.13 from (5.1) to (5.2). Gain K for the constant impedance in Fig. 5.13 is

$$\frac{\textbf{[PL(7); PL(9)-i[QL(7); QL(9)]]}}{\textbf{(abs[vol(7); vol(9)])}}$$

Then build four new Simulink models. The first one for a constant power load with a delay block named `'delay_block_fourmach_withcontrol_sys_base_ConstP'` and the second one with the same load but an IC block, named `'IC_block_fourmach_withcontrol_sys_base_ConstP'`. The third model is for the constant impedance load with a delay block `'delay_block_fourmach_withcontrol_sys_base_ConstZ'`. The fourth model is the constant impedance load with the IC block implementation, `'IC_block_fourmach_withcontrol_sys_base_ConstZ'`. Finally, we will refer to the original Simulink model with implicit load representation as `'Implicit_method'`, and the model should be saved according to this name.

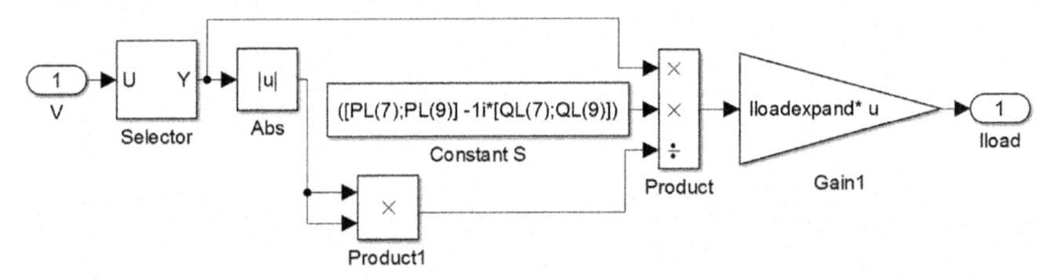

Figure 5.12 Explicit load representation of constant power load.

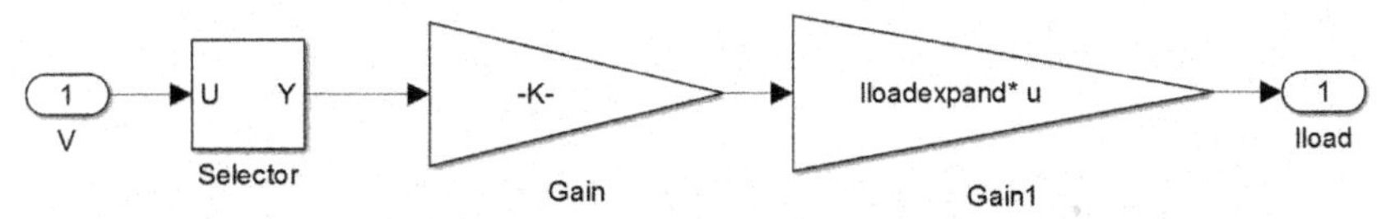

Figure 5.13 Explicit load representation of constant impedance load.

In total, we have developed seven different load models in this chapter — two implementation methods for each type (constant Z, I or P) and the additional implicit implementation for the Z-load. Script 5.7 allows us to view all seven loads on one eigenvalue plot. For clarity, we will look at the poles of the constant power and constant impedance load in turn, before turning to the overall comparison of poles.

Fig. 5.14 shows the relevant poles of the four-machine system when the loads (both P and Q) are modelled as constant power loads, and the algebraic loop is broken once with an IC block and once with a delay block. As was the case for the constant current load, we can see that the poles agree, independent of the method chosen to break the loop.

Fig. 5.15 shows the poles when the system has only constant impedance loads. The implicit method is the best method to model such a system, as it does not contain any algebraic loops, which leads to an exact result.

```
xstates= [s_Eqd; s_delta; s_Psi1d; s_Psi2q; s_Edd; s_Efd./s_Ka; s_ws;
s_ws;s_ws;s_ws; s_Tm];

%Constant I
sys = linmod('IC_block_fourmach_withcontrol_sys_base',xstates,[],...
[1e-50.001 0])
eig(sys.a)
scatter(real(eig(sys.a)), imag(eig(sys.a)),'ro');
hold on
sys = linmod('delay_block_fourmach_withcontrol_sys_base')
eig(sys.a)
scatter(real(eig(sys.a)), imag(eig(sys.a)),'r*')

%Constant P
sys =
linmod('IC_block_fourmach_withcontrol_sys_base_ConstP',xstates,[],...
[1e-50.001 0])
eig(sys.a)
scatter(real(eig(sys.a)), imag(eig(sys.a)),'bo');

sys = linmod('delay_block_fourmach_withcontrol_sys_base_ConstP')
eig(sys.a)
scatter(real(eig(sys.a)), imag(eig(sys.a)),'b*')

%Constant Z
sys =
linmod('IC_block_fourmach_withcontrol_sys_base_ConstZ',xstates,[],...
[1e-50.001 0])
eig(sys.a)
scatter(real(eig(sys.a)), imag(eig(sys.a)),'ko');
sys = linmod('delay_block_fourmach_withcontrol_sys_base_ConstZ')
eig(sys.a)
scatter(real(eig(sys.a)), imag(eig(sys.a)),'k*')
sys = linmod('Implicit_method')
eig(sys.a)
scatter(real(eig(sys.a)), imag(eig(sys.a)),'k^')
grid on
```

Script 5.7 Linearization of all seven presented load models.

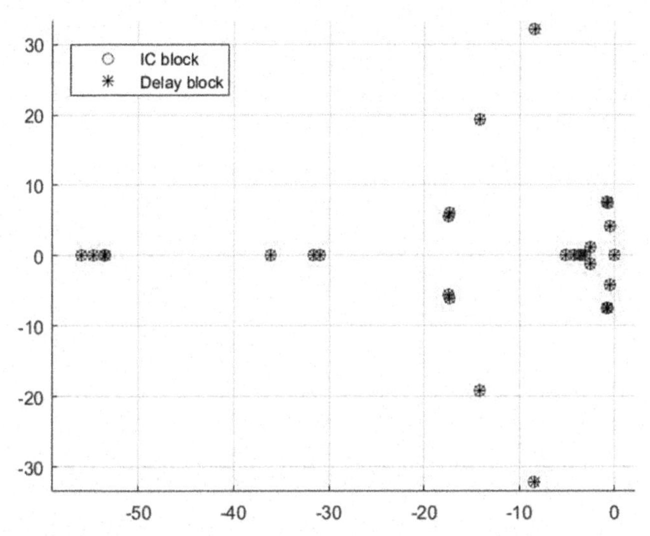

Figure 5.14 Zoomed view of relevant poles of constant power load with delay block and initial condition (IC) block.

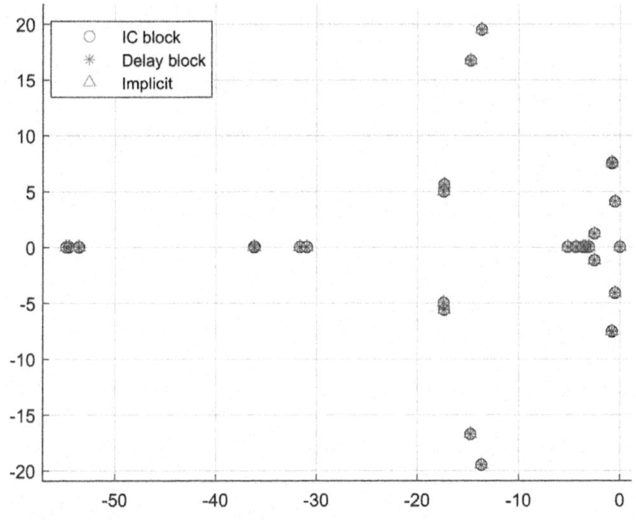

Figure 5.15 Zoomed view of relevant poles of constant impedance load with delay block, IC block and implicit method.

However, using the explicit method with the IC block or the delay block gives results that are extremely close to the exact (implicit) method. This result is significant because it can provide confidence in both methods, which may be used for the other load types.

Now we have the eigenvalue plots of all three types of loads (constant Z, I and P), we can compare all the results on one eigenvalue plot to see how different load types impact the poles. Fig. 5.16 shows the zoomed view of all relevant poles. It can be seen that many poles are very similar between all load types, with markers overlapping in the plot. Poles with higher frequencies, above 10 rad/sec, show a significant difference in mode locations. We now zoom into the group of poles located closest to the right-hand side, as shown in Fig. 5.17. There is very little difference between any poles within this group.

After this exhaustive discussion of the eigenvalues, let us follow by inspecting the difference that can be seen in time domain analysis, through the change in load type. Fig. 5.18 shows the voltage response of one of the two load buses (Bus 7), when the generator excitation voltage reference is increased by 1% after 1 s of steady-state simulation. The difference is clearly visible, and in this case, all three scenarios settle to a new steady state. Have a go at creating this result yourself, you can use the 'to file' block under sinks, to conveniently capture and save time domain simulations in Simulink, and you can use the selector block to only capture the data for the buses you are interested in. For creating the disturbance, replace the constant block at your excitation control with a 'step' available under sources, and increase your reference value by 1% at 1 s. If you set the filenames, for the three 'to file' blocks as ConstP, ConstI and ConstZ, you can use Script 5.8 to achieve the same plot as Fig. 5.18.

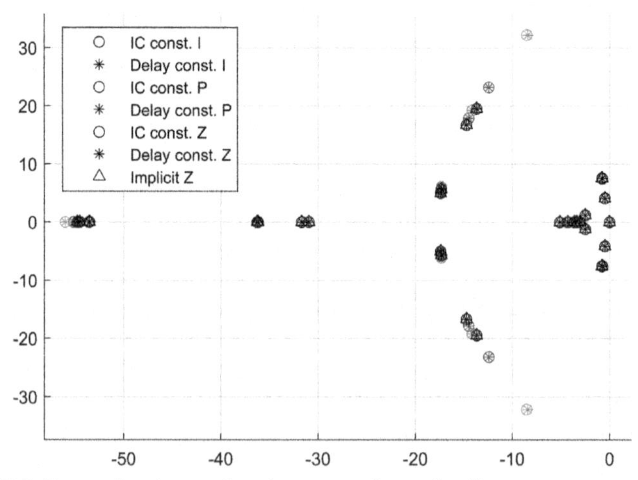

Figure 5.16 Zoomed view of relevant poles of all seven presented load representations.

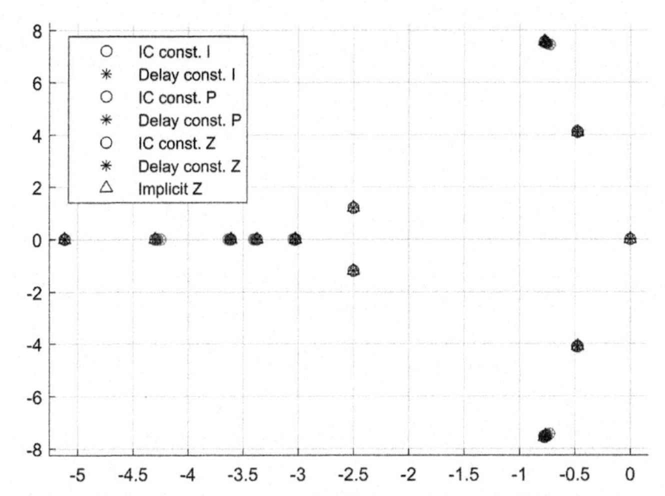

Figure 5.17 Further zoomed in view of relevant poles of all seven presented load representations.

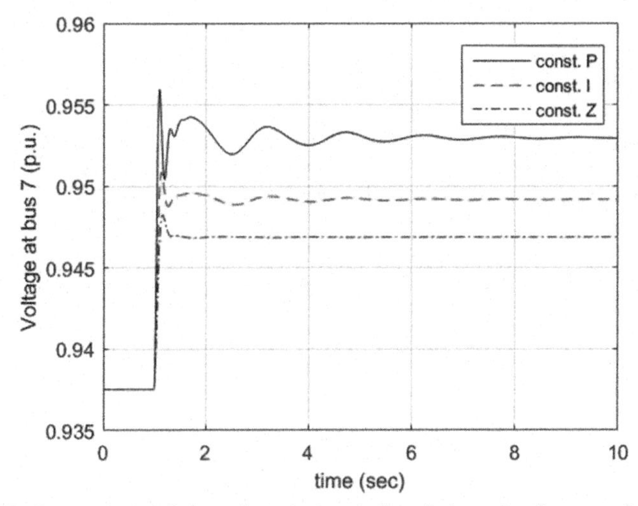

Figure 5.18 Comparison of time domain result for all three load types, showing bus voltage at Bus 7 (load bus) when the voltage reference to the generators' excitation control is increased by 1% at time 1 s.

```
load('ConstP.mat')
plot (ans, 'k')
hold on
load('ConstI.mat')
plot (ans,'r --')
load('ConstZ.mat')
plot (ans,'b -.')
grid on
```

Script 5.8 Script for time domain plot.

5.6 Conclusion of ZIP load modelling

In this chapter, we have discussed that pure Z-load models can easily be incorporated in power system studies, and dynamic load models would neither incur algebraic loop issues. Other algebraic load models, such as ZIP loads, do introduce algebraic loops. In terms of Simulink simulations, we can handle this using IC blocks or delay blocks, and a higher level of confidence and certainty may even be gained when both methods are applied and contrasted. The developed ZIP load models were used in a four-generator system, and the difference in poles and time domain response with a change of load behaviour has been analyzed.

Acknowledgement

We wish to thank Husni Rois Ali, who first became aware of the special procedure required in the use of IC blocks for linearization, and Jeff Homer from the MathWorks Technical Support Department, who suggested a suitable work around.

References

Asres, M., Girmay, A., Camarda, C., Tesfamariam, G., 2019. Non-intrusive load composition estimation from aggregate ZIP load models using machine learning. Int. J. Electr. Power Energy Syst. 105, 191–200. Elsevier.
Kundur, P., 1994. Power System Stability and Control. McGraw-Hill.
Zhao, J., Wang, Z., Wang, J., July 2018. Robust time-varying load modeling for conservation voltage reduction assessment. IEEE Trans. Smart Grid 9 (4).

CHAPTER SIX

Wind turbine generator modelling

This chapter describes the procedure to construct and run a wind farm simulation program considering doubly fed induction generator (DFIG)– and permanent magnet synchronous generator (PMSG)–based wind turbine technologies. By the end of the chapter, the readers will acquire the competence to simulate a wind turbine generator (WTG) connected to an infinite bus in Matlab/Simulink software. The chapter also includes an example to build a model of a wind farm containing mix of DFIG- and PMSG-type WTGs. It is advised to build the Simulink model while reading the chapter and use the example results provided to validate their model.

The chapter commences with modelling of DFIG-based WTG, which is divided into seven component models such as models of network, turbine, generator, filter, converter capacitor, machine-side converter (MSC) and grid-side converter (GSC). Equations describing each component model, figures for their Simulink model and associated Matlab programs are presented. Modelling of the DFIG is followed with simulation results on a single machine infinite bus (SMIB) test system. Dynamic simulation results and modal analysis results are presented for comparison. Programming of the PMSG-based WTG model is discussed later, which utilizes many of the subsystems of the DFIG system. At the end, the models developed are used to build a simulation program for a wind farm containing both types of WTGs.

6.1 Introduction

Wind is the fastest growing renewable energy source in the world. As the cost of electricity from wind, especially offshore wind farms, is decreasing, wind farms will contribute to a major portion of electricity generated in many electricity grids. Their characteristics will significantly influence the stability and operation of the grid soon. So, the modelling and analysis of wind turbines will play an important role in power system studies.

The early generation WTGs is fixed speed machines using induction generators. The rotor speed is fixed relative to grid frequency and independent of the wind speed. The output control is achieved using pitch and stall control mechanisms. The use of DFIG and power electronics converters

Simulation of Power System with Renewables
ISBN: 978-0-12-811187-1
https://doi.org/10.1016/B978-0-12-811187-1.00006-8
© 2020 Elsevier Inc.
All rights reserved.

made variable speed operation possible, allowing greater power extraction under wide range of wind speeds. More recently, full converter–based WTGs have been developed, whose generator is isolated from grid through a power electronics converter making the generator rotor speed independent of the grid frequency. PMSGs are widely used in the full converter WTG configuration. The DFIG- and full converter–based WTGs offer reactive power and voltage control capabilities. This chapter will explain the modelling of DFIG- and PMSG-type WTGs.

Note: In the model developed in Chapter 3, variables of synchronous machine have 's_' prefix. Similarly, the DFIG and PMSG variables discussed in this chapter use 'd_' and 'p_' prefixes, respectively.

6.2 Building blocks of DFIG-SMIB simulation model

Fig. 6.1 shows building blocks of a DFIG-SMIB system, which consists of a wind turbine, DFIG and network. The stator windings of the DFIG are fed directly from the grid, whereas the rotor windings are fed through a back-to-back converter, which is connected to the grid through a filter. The frequency and magnitude of the rotor voltage is controlled through the back-to-back converter, which can carry up to 30% of the active power output depending on the rotor speed. The filter is used to remove switching frequency harmonic components.

For modelling purposes, the system is divided into seven blocks, namely turbine, network, generator, MSC, back-to-back capacitor (B2BC), GSC and filter. The exchange of data between these main blocks is indicated using arrows in the figure. For example, inputs to the generator model are rotor speed from the turbine block and generator bus voltage from the network block. The generator model sends electrical torque to the turbine model and output current to the network model. The division into

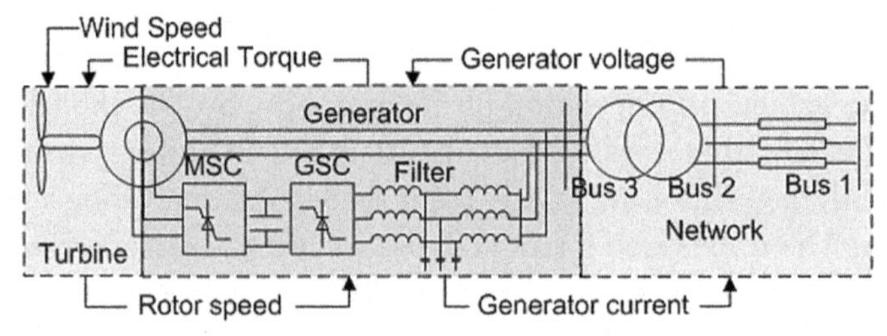

Figure 6.1 Schematic diagram of a single machine infinite bus (SMIB) system using doubly fed induction generator (DFIG)–type wind turbine generator.

submodels helps to develop and test individual blocks separately before integrating them to a DFIG-SMIB system simulation model. Readers are encouraged to develop the model and test as they go through this chapter to achieve the best outcome. The Script 6.1 contains parameters for the system at one operating condition, which can be used to test the individual blocks and the complete DFIG-SMIB system. Before proceeding to next

```
%Values associated with network
Znet = 0.0472 + 1i*0.4700;  d_Vinf = 1;   Dmachs = 3;
% Values associated with generator
d_Lm = 4;            d_Xm = d_Lm;       d_Rs = 0.005; d_Rr = 0.0055;
d_Lss = 4.04;        d_Lrr = 4.0602; d_kopt = 1;    d_ktg = 0.3;
d_ctg = 0.01;        d_Ht = 4;          d_Hg = 0.4;

d_Ls_d = d_Lss - (d_Lm^2/d_Lrr);
d_Kmrr = d_Lm/d_Lrr;
d_R2 = d_Kmrr^2*d_Rr;
d_R1 = d_Rs + d_R2;
d_Tr = d_Lrr/d_Rr;

Vdfig = 0.9794 + 1i*0.3983;
d_vsq = real(Vdfig);          d_vsd = imag(Vdfig);
d_Theta = angle(Vdfig);

d_isq = 0.8544;        d_isd = 0.2454;
d_irq =  -0.9629;      d_ird = -0.0020;
d_vrq =  0.0357;       d_vrd = 0.0154;

% Values associated with Filter
d_Ri= 0.0;    d_Rg= 0.0;     d_Rc= 0.7333;     d_Li= 0.1667;
d_Lg= 0.0033;     d_Cf = 0.0150;
d_iiq   = -0.0361;     d_iid   = 0.0024;
d_igq   = -0.0303;     d_igd   = -0.0123;
d_viq   = 0.9790;    d_vid   = 0.3922;
d_vcq   = 0.9837;    d_vcd   = 0.3874;


% Values associated with converters
d_Cdc = 2; d_VDC = 1.5;   d_Qfilter = 0;    d_Qs = 0.1;

d_MSC_IL1_kp = -0.23;  d_MSC_IL1_ki = -3;  d_MSC_IL1_iv = 0.0389;
d_MSC_IL2_kp = -0.23;  d_MSC_IL2_ki = -3;  d_MSC_IL2_iv = 0.00078156;
d_MSC_OL1_kp = 0;      d_MSC_OL1_ki = -60; d_MSC_OL1_iv = -0.8927;
d_MSC_OL2_kp = 0;      d_MSC_OL2_ki = 90;  d_MSC_OL2_iv = 0.3610;

d_GSC_IL1_kp = 0.3; d_GSC_IL1_ki = 200;  d_GSC_IL1_iv = 1.0546;
d_GSC_IL2_kp = 0.3; d_GSC_IL2_ki = 200;  d_GSC_IL2_iv = -0.0055;
d_GSC_OL1_kp = -22; d_GSC_OL1_ki = -870; d_GSC_OL1_iv = -0.0327;
d_GSC_OL2_kp = 0;   d_GSC_OL2_ki = -60;  d_GSC_OL2_iv = 0;


% Values associated with turbine
rho = 1.225;
d_wtrated = 3.0337; d_wt = 0.9688; d_wg = 0.9688;
d_bl = 40.05; d_Lambda = 8.1;
d_Beta = 0; d_vw = 14.5316; d_Ts = 0.9385;
```

Script 6.1 Test values for the doubly fed induction generator (DFIG)—single machine infinite bus (SMIB) model. Save as test_dfig.m.

section, copy this script and save it as *test_dfig.m*. Section 6.4 discusses the method to find these values for different operating conditions.

6.2.1 Network

In Chapter 3, we have developed a simulation program for a synchronous machine connected to an infinite bus. The Simulink block for the network developed in Section 3.5 has been reused for the DFIG-SMIB model. However, we will use a different bus and line matrix given in Script 6.2 to match with the new network in Fig. 6.1. The script is used to calculate the network impedance. These data refer to a 5 MW WTG on a 5 MVA base.

```
% This script contains the bus and line matrix for an example three
% bus
% network shown in Figure 6.1. The WTG is connected at Bus 3 and Bus
% 1 is the infinite bus. The WTG output is fixed and hence bus 3 is
% defined as a load bus. Ensure that functions,
% form_Ymatrix.m and power_flow.m, given in Chapter 2 are saved in
% the current working folder. The functions are used to obtain the
% power flow and admittance matrix. The network impedance to use in
% the network block of Simulink program is calculated from the
% admittance matrix.

clear all
%**** Part 1: Power Flow, Calculation of Network Impedance    *******
% bus data format
% bus: number, voltage(pu), angle(degree), p_gen(pu), q_gen(pu),
%      p_load(pu), q_load(pu),G-shunt (pu), B shunt (pu); bus_type
%      bus_type - 1, swing bus
%               - 2, generator bus (PV bus)
%               - 3, load bus (PQ bus)

bus = [...
    01 1.05  0.00  1.00  0.00  2.00   0.30  0.00  0.00  1;
    02 1.00  0.00  0.00  0.00  0.00   0.00  0.00  0.00  3;
    03 1.00  0.00  0.80  0.10  0.00   0.00  0.00  0.00  3];

% line data format
% line: from bus, to bus, resistance(pu), reactance(pu),
%       line charging(pu), tap ratio

line = [
    01  02 0.037  0.37  0.001 0.0 0.0;
    02  03 0.010  0.10  0.000 0.0 0.0];

Y = form_Ymatrix(bus,line);
[bus_sln, flow] = power_flow(Y, bus, line);
% Generator bus
Gen_Bus = 3;
% Infinite bus voltage
vinf = bus_sln(1,2).*exp(1i*bus_sln(1,3)*pi/180);
% Generator bus voltage
Vg = bus_sln(Gen_Bus,2).*exp(1i*bus_sln(Gen_Bus,3)*pi/180);
% Current injection from generator
d_ig = conj((bus_sln(Gen_Bus, 4)+1i*bus_sln(Gen_Bus, 5))/Vg);

Znet = (Vg - vinf)/d_ig;   % Network Impedance
```

Script 6.2 Network impedance calculation.

The bus matrix has three rows: one of them represents a slack bus and other two PQ buses. Bus 1 is the infinite bus and Bus 3 is where the WTG is connected. The WTG's active and reactive powers are specified in row 3, which can be changed to simulate different operating conditions. Line matrix has two rows indicating a transformer and infinite bus impedance. Now one can develop a network model as shown in Fig. 3.9 and Fig. 3.10 in Section 3.5.

6.2.2 Wind turbine model

The wind turbine model relates the wind velocity with generator rotor speed. A typical turbine configuration has three blades that capture wind energy and rotate a generator rotor. Depending on the generator speed requirements, a gearbox may be placed between the turbine and generator. As shown in Fig. 6.1, the turbine model block receives wind speed and electrical torque inputs and provides generator speed as output.

The turbine model is divided into two submodels. (1) An aerodynamic model converting wind energy to mechanical torque and (2) a drive train model relating the mechanical torque, electrical torque input from generator block and generator speed output.

6.2.2.1 Wind turbine aerodynamic modelling

The wind turbine aerodynamic model relates wind speed with turbine mechanical output. Let the example turbine have three blades of length R (40.05 m). The equation for mechanical power output is given by (6.1)

$$P_t = 0.5\rho\pi R^2 C_p(\beta, \lambda)v_w^3[W] \tag{6.1}$$

where ρ is air density $= 1.225\ \text{kg/m}^3$, $\pi = 3.1416$, v_w is the wind velocity in m/s and $C_p(\beta,\ \lambda)$ is the power coefficient of the blades. The theoretical maximum value of C_p is limited to 0.59, which is called Betz limit.

Eq. (6.1) shows that for a wind speed v_w, the turbine output depends on the C_p coefficient, which is a function of two parameters, β and λ. C_p represents the portion of wind power that can be extracted by the turbine. Usually manufacturers, using field-testing data, specify the performance curve for their turbine. However, for academic studies, numerical approximate equations can be used. This chapter will use one such equation (Mei, 2008) given in (6.2).

$$C_p(\beta,\ \lambda) = 0.5176\left(\frac{116}{\lambda + 0.08\beta} - \frac{4.06}{1 + \beta^3} - 0.4\beta - 5\right)e^{\left(\frac{-21}{\lambda+0.08\beta} + \frac{0.735}{1+\beta^3}\right)} + 0.0068\lambda \tag{6.2}$$

where β is the pitch angle of the blade and λ is the tip speed ratio. The blades of the turbine can be turned in and out of the wind to control the turbine speed. The **pitch angle** refers to the angle of rotation of the blade on its longitudinal axis. For the wind turbine data used in this chapter, the pitch angle varies from 0 to 23 degrees; the pitch angle is zero degrees when the blades are facing the wind extracting maximum energy. The **tip speed ratio** is the ratio between the speed of the tip of a turbine blade and the wind speed, i.e., $\lambda = \omega_t R / v_w$, where ω_t is the turbine rotational speed in rad/sec.

Plots of C_p coefficient for different values of pitch angles (0, 4 and 8 degrees) are shown in Fig. 6.2. It is clear that C_p decreases as the blades are turned away from the wind by increasing the pitch angle. For a given wind speed v_w and pitch angle β, C_p (and hence turbine output power P_t) varies with the tip speed ratio λ. The tip speed ratio can be controlled by controlling the turbine speed ω_t.

Yaw control: This is another important mechanism to orient the turbine into the wind by turning the nacelle of the wind turbine. In older WTGs, the control was used to regulate the output. However, modern WTGs do not use Yaw control for power regulation. This control is not discussed

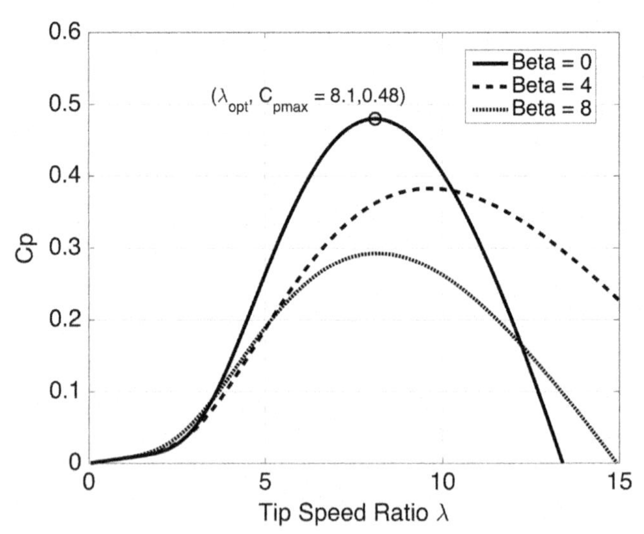

Figure 6.2 Plot of turbine performance coefficient versus tip speed ratio for different pitch angles.

further in this chapter. A simplified assumption is made such that the yaw control keeps the turbine aligned with wind direction always.

The maximum value for C_p is achieved for $\beta = 0$ when the blades are turned into the wind. A plot of C_p against λ for $\beta = 0$ is shown by a solid line in Fig. 6.2. For the given turbine parameters, the maximum value of C_p called $C_{pmax} = 0.48$ occurs at $\lambda_{opt} = 8.1$, which is the optimum value of tip speed ratio at $\beta = 0$. Substituting $C_p = C_{pmax}$ and $P_t = 5$ MW in (6.1), the wind speed required to produce the rated output at $\beta = 0$ is obtained as 15 m/s. This is the **rated wind speed** of the turbine. Above the rated wind speed, the pitch angle of the turbine is increased to reduce C_p to limit output to 5 MW. The operating region at and above rated wind speed is called **rated wind speed region**. Below the rated wind speed, the wind turbine output will be less than rated output, and the region is called **subrated operating region**.

In the subrated region, the maximum power from wind is extracted by setting pitch angle $\beta = 0$ and tip speed ratio $\lambda = \lambda_{opt}$. The later condition is met by adjusting the rotational speed of the turbine $\omega_t = \lambda_{opt} v_w / R$. However, turbine and generator designs restrict deviation in turbine rotational speed. For example, in case of a DFIG, rotor slip must be within $\pm 30\%$. The performance coefficient will not be constant for the entire range of subrated operating region, especially close to **cut in wind speed**, which is the minimum wind speed required for a turbine to start producing useful output. However, for simplicity, assume $C_p = C_{pmax}$ at the subrated operating region (*This assumption results in constant tip speed ratio (8.1 at $\beta = 0$) under the subrated operating region is irrespective of wind speed. This means a large deviation in the rotor speed and slip, which is not acceptable. Hence, for the given parameters, the assumption $C_p = C_{pmax}$ makes the simulation model suitable for a limited range of wind speed. Readers must take note of this point. This issue can be fixed by using more accurate turbine parameters*). Above the rated wind speed, β and λ are varied to control C_p and hence limit the turbine output P_t to the rated value. The dashed and dotted plots in Fig. 6.2 show C_p curves for $\beta = 4$ degrees and $\beta = 8$ degrees, respectively. The pitch angle control is used to decrease C_p at above rated wind speed and hence controls turbine output. However, beyond a certain wind speed, called **cut out wind speed**, the turbine shuts down to protect itself from mechanical damage due to the strong wind. A plot between wind speed and turbine output for a typical WTG is shown in Fig. 6.3 The cut in wind speed, rated wind speed and cut out wind speed are clearly marked.

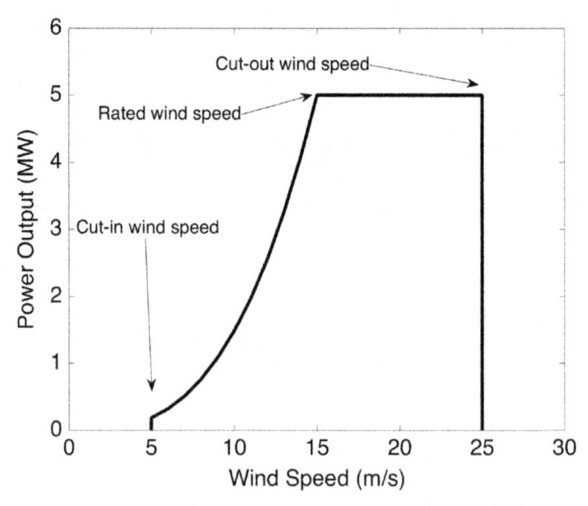

Figure 6.3 Turbine output versus wind speed plot.

6.2.2.1.1 Simulink representation of turbine model

Open a new Simulink model file and add a subsystem block. The subsystem block can be found in the 'commonly used blocks' section. Rename the subsystem block to Cp_Calc. Open the Cp_Calc block and implement the turbine performance coefficient equation in (6.2) as shown in Fig. 6.4. Save the model as Turbine_model.slx. To test the model, add two constant blocks and a display block as shown in Fig. 6.5. Now run test_dfig.m and Turbine_model.slx and verify the value in the display block.

Having implemented the Cp_Calc block, it is time to relate the wind speed input to turbine torque output. Append the model with additional elements to create the Simulink representation of the turbine aerodynamic model as shown in Fig. 6.6. The tip speed ratio is calculated using the equation $\lambda = \frac{\omega_t R}{v_w} = \frac{\omega_{t\ pu}\ \omega_{t\,\mathrm{rated}} R}{v_w}$, where the turbine rated speed is $\omega_{t\,\mathrm{rated}} = 3.0337$ rad/sec. The variable d_wt in the *constant* block represents pu speed. The *Gain3* block is used to convert the actual turbine power to the per unit value by dividing the actual value by the machine base (d_base). The output of the *Gain3* block is turbine mechanical power in per unit, and dividing it by the p.u. turbine speed produces turbine mechanical torque. Now run test_dfig.m and the Simulink model. Once model is validated, select all the components except the constant blocks and display block, click right mouse button and then select the *Create Subsystem From Selection* option to make a subsystem. Name the new subsystem *Turbine_Aero*. Refer to Fig. 6.9 for guidance.

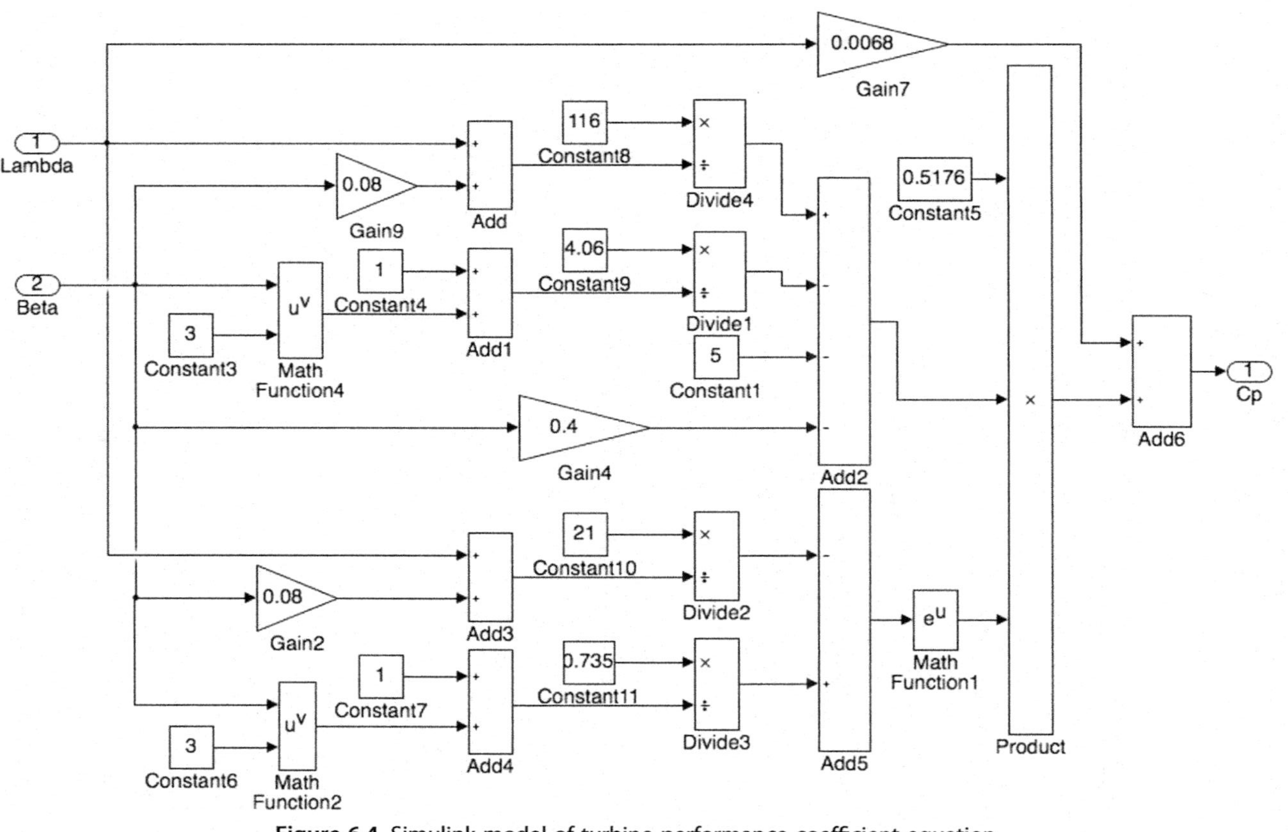

Figure 6.4 Simulink model of turbine performance coefficient equation.

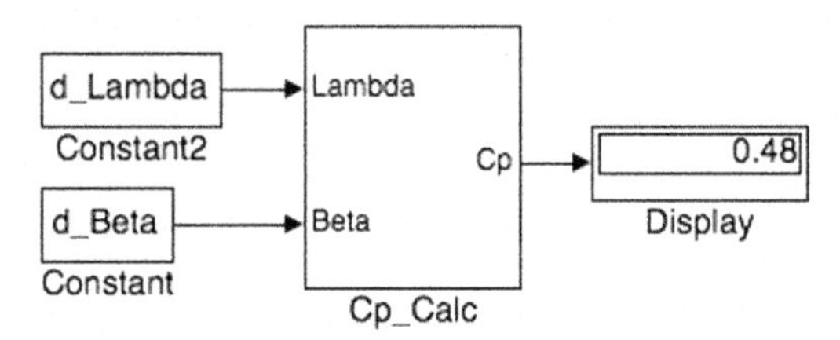

Figure 6.5 Simulink subsystem for turbine performance coefficient equation.

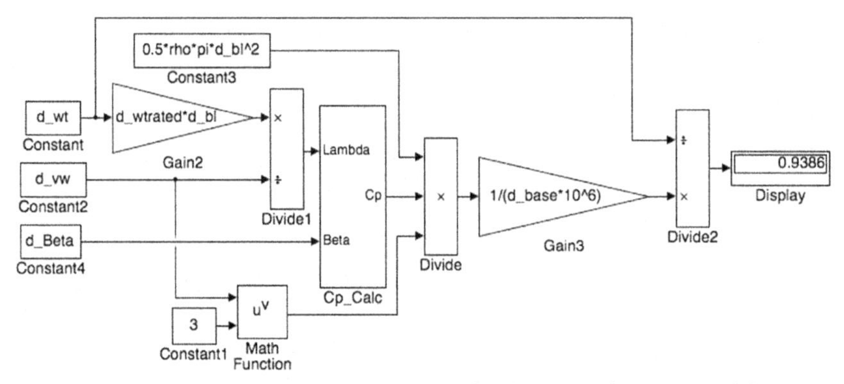

Figure 6.6 Simulation representation of turbine aerodynamic model.

6.2.2.2 Turbine generator mechanical drive train model

The second part of the wind turbine model is the drive train model relating the mechanical torque, electrical torque from the generator block and generator speed output. It represents the turbine, gearbox and generator masses. Models of varying complexity are reported in the literature, as already discussed in Chapter 3. For the simulations presented in this chapter, the drive train is modelled using two masses coupled through a shaft as shown in Fig. 6.7. Eqs. (6.3–6.6) are used to represent the model. The parameters and their values are listed in Table 6.1. The variable name used in the program is shown in brackets under the column symbol.

$$\frac{d}{dt}\omega_g = \frac{1}{2H_g}(T_s - T_g) \tag{6.3}$$

$$T_s = k_{tg}\theta_{tg} + c_{tg}\frac{d}{dt}\theta_{tg} \tag{6.4}$$

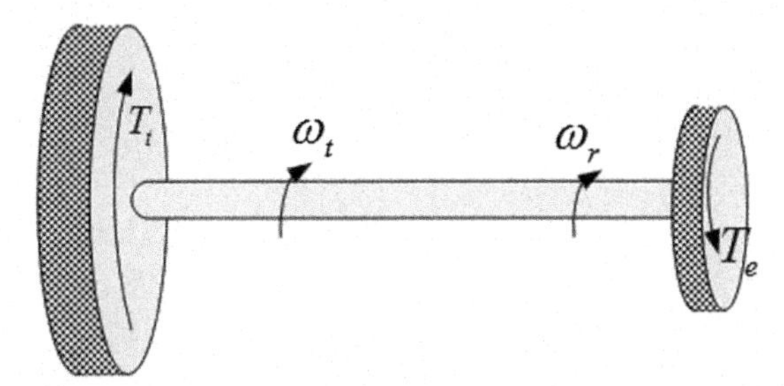

Figure 6.7 A two-mass drive train model of wind turbine.

$$\frac{d}{dt}\theta_{tg} = \omega_{elB}(\omega_t - \omega_g) \tag{6.5}$$

$$\frac{d}{dt}\omega_t = \frac{1}{2H_t}(T_t - T_s) \tag{6.6}$$

where T, ω and θ_{tg} represent torque, rotational speed and shaft twist angle. The subscripts g, s and t refer to generator, shaft and turbine, respectively.

The system of equations requires two inputs: turbine mechanical torque T_t and the electrical torque produced by the generator coupled to the turbine T_g. The speed of turbine ω_t and generator ω_g are outputs obtained from the system of equations. A Simulink representation of the equations is shown in Fig. 6.8. You are advised to compare the figure and equations and build the model yourself. As explained in Chapter 1, the initial values of state variables are specified in the block parameters dialogue box of the *Integration* block under *initial condition*. The initial values are calculated using a separate Matlab script file that will be discussed in Section 6.4. In this book, the variable for the initial value for each state variable is used as the name of the integration block. For example, in Fig. 6.8, d_Ts is the name of an integration block. Double click the block to open its block parameter dialogue

Table 6.1 Drive train parameters of the wind turbine.

Parameter	Symbol (variable name)	Value
Turbine inertia	H_t (d_Ht)	4 s
Generator inertia	H_g (d_Hg)	0.4 s
Drive train shaft stiffness	k_{tg} (d_ktg)	0.3 pu/el.rad
Drive train damping coefficient	c_{tg} (d_ctg)	0.01 pu.s/el.rad
Electrical base speed	ω_{elB} (d_welb)	$2\pi 50$ rad/s

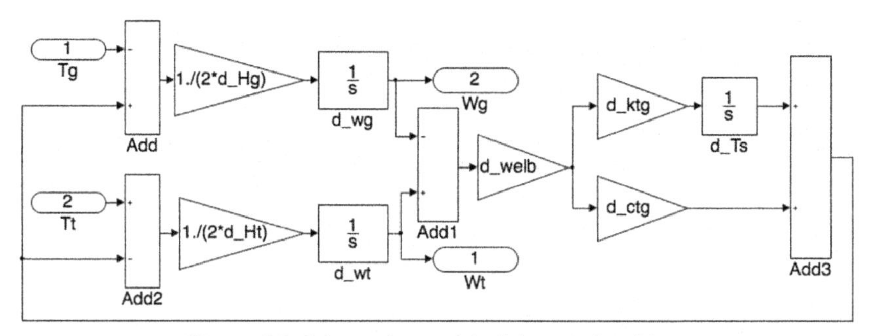

Figure 6.8 Drive train model of the wind turbine.

box and input d_Ts in the initial condition section of the dialogue box. Now convert the model to a subsystem named *Drive_Train* and connect it to the *Turbine_Aero* subsystem as shown in Fig. 6.9.

The integrated model takes wind speed and generator electrical torque as input and generator speed as output. Test the model after loading the variables from dfig_test.m. The turbine model of the DFIG-SMIB system is ready now.

6.1.3 Doubly fed induction generator

So far, the wind turbine and network representation have been discussed. The missing link is a generator model that takes generator rotor speed (ω_g) from the turbine model and generator bus voltage (v_g) from the network model as inputs and produces generator output current (i_g) and electrical torque (T_g) as outputs.

As the name suggests, a DFIG is an induction generator having two feeds: one through to the stator and another one through to the rotor windings. The stator is connected to a three-phase voltage at grid frequency (f_s = 50 or 60 Hz). The stator windings are designed such that the resulting stator currents produce a rotating magnet field in the air gap with angular

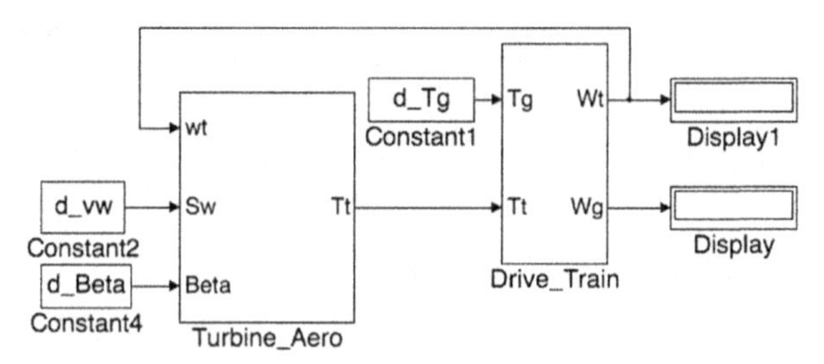

Figure 6.9 Turbine model of the single machine infinite bus (SMIB) system.

speed, $\omega_s = 2*pi*f_s$. To transfer energy from the rotor windings to the stator windings, there must be a synchronously rotating magnetic field produced by the rotor windings in the air gap. That means if the turbine drives the rotor at a speed $\omega_g = \omega_s$, a direct current through the rotor windings can produce a magnetic field rotating synchronously with the rotating field produced by the stator windings. But for a wind turbine, generator speed varies as the wind speed fluctuates. Suppose the turbine drives the rotor at a speed slightly higher or lower than the synchronous speed ($\omega_g \neq \omega_s$), an alternating rotor current at slip frequency $f_r = (\omega_s - \omega_g)f_s$ is required to generate a rotor field at the synchronous speed. This variable frequency feeding to rotor windings helps the DFIG-based WTG to continuously generate output even when the rotor speed varies due to changes in wind speed. This is the main feature that made the DFIG-WTG attractive to wind farm developers. As shown in Fig. 6.1, a back-to-back converter is used to produce the rotor voltage at variable frequency and magnitude. The full converter–based WTG discussed later has more flexibility as its converter isolates the generator from grid.

The model of the back-to-back converter is discussed in Sections 6.5–6.7. For the time being, let us develop a model of the DFIG assuming that the rotor voltage is supplied by an ideal voltage source as shown in Fig. 6.10. $v_{abc,s}$ and $v_{abc,r}$ are the three-phase voltage applied to stator and rotor windings, respectively. The differential equations representing the machine are presented in a d-q reference frame in which the q-axis is aligned with the infinite bus a-phase voltage. We have discussed the dq-reference frame concept in Chapter 3.

Eqs. (6.7–6.23) represent the differential and algebraic equations governing the dynamics of the DFIG (Mei, 2008; Abad et al., 2011; Wu et al., 2011) shown in Fig. 6.10. The definitions of the variables used in the model and their values are listed in Table 6.2.

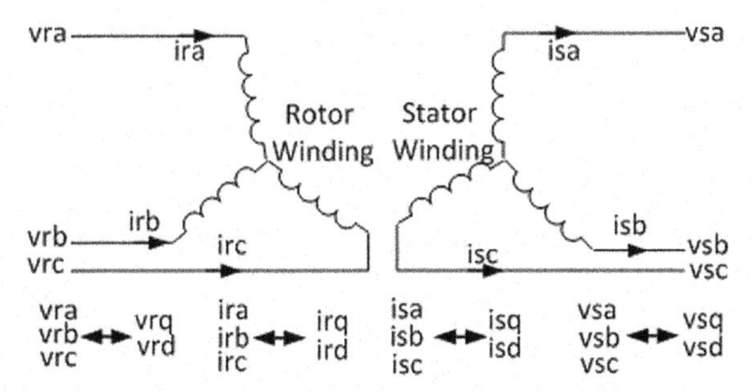

Figure 6.10 Simplified representation of a doubly fed induction generator (DFIG).

Table 6.2 Parameters of doubly fed induction generator.

Parameter	Symbol (variable name)	Value
Mutual inductance	L_m (d_Lm)	4 pu
Stator inductance	L_s (d_Ls)	4.04 pu
Rotor inductance	L_r (d_Lr)	4.0602 pu
Stator resistance	R_s (d_Rs)	0.005 pu
Rotor resistance	R_r (d_Rr)	0.0055
	$K_{mrr} = L_m/L_r$	
	$R_2 = K_{mrr}^2 R_r$	
	$R_1 = R_2 + R_s$	
	$L_s^{'} = L_s - L_m K_{mrr}$	
	$T_r = L_r/R_r$	
	$X_m = L_m$	

Stator currents (i_{sq}, i_{sd}) are

$$\frac{L_s'}{\omega_{elB}}\frac{d}{dt}i_{sd} = -\omega_s L_s' i_{sq} - R_1 i_{sd} + \frac{e_{sq}'}{\omega_s T_r} + \frac{\omega_g e_{sd}'}{\omega_s} - v_{sd} + K_{mrr} v_{rd} \tag{6.7}$$

$$\frac{L_s'}{\omega_{elB}}\frac{d}{dt}i_{sq} = -R_1 i_{sq} + \omega_s L_s' i_{sd} + \frac{\omega_g e_{sq}'}{\omega_s} - \frac{e_{sd}'}{\omega_s T_r} - v_{sq} + K_{mrr} v_{rq} \tag{6.8}$$

Voltages behind transient impedance (e_{sq}', e_{sd}') are

$$\frac{1}{\omega_s \omega_{elB}}\frac{d}{dt}e_{sd}' = -R_2 i_{sq} - \left(1 - \frac{\omega_g}{\omega_s}\right)e_{sq}' - \frac{e_{sd}'}{\omega_s T_r} + K_{mrr} v_{rq} \tag{6.9}$$

$$\frac{1}{\omega_s \omega_{elB}}\frac{d}{dt}e_{sq}' = R_2 i_{sd} - \frac{e_{sq}'}{\omega_s T_r} + \left(1 - \frac{\omega_g}{\omega_s}\right)e_{sd}' - K_{mrr} v_{rd} \tag{6.10}$$

where, e_{sq}' and e_{sd}' are related to rotor flux ψ_{rq} and ψ_{rd}, respectively, by

$$e_{sq}' = K_{mrr}\omega_s \psi_{rd} \tag{6.11}$$

$$e_{sd}' = -K_{mrr}\omega_s \psi_{rq} \tag{6.12}$$

The stator and rotor fluxes are given by

$$\psi_{sq} = L_s i_{sq} + L_m i_{rq} \tag{6.13}$$

$$\psi_{sd} = L_s i_{sd} + L_m i_{rd} \tag{6.14}$$

$$\psi_{rq} = L_r i_{rq} + L_m i_{sq} \tag{6.15}$$

$$\psi_{rd} = L_r i_{rd} + L_m i_{sd} \tag{6.16}$$

Rotor currents i_{rq} and i_{rd} are

$$i_{rq} = -\left(\frac{e'_{sd}}{X_m}\right) - K_{mrr} i_{sq} \tag{6.17}$$

$$i_{rd} = \left(\frac{e'_{sq}}{X_m}\right) - K_{mrr} i_{sd} \tag{6.18}$$

Stator active power P_s, rotor active power P_r, stator reactive power Q_s, rotor reactive power Q_r and electric torque T_g are given by

$$P_s = v_{sq} i_{sq} + v_{sd} i_{sd} \tag{6.19}$$

$$P_r = v_{rq} i_{rq} + v_{rd} i_{rd} \tag{6.20}$$

$$Q_s = -v_{sq} i_{sd} + v_{sd} i_{sq} \tag{6.21}$$

$$Q_r = -v_{rq} i_{rd} + v_{rd} i_{rq} \tag{6.22}$$

$$T_g = L_m (i_{sq} i_{rd} - i_{sd} i_{rq}) \tag{6.23}$$

Assume that stator voltage (v_{sq}, v_{sd}), rotor voltage (v_{rq}, v_{rd}) and rotor speed ω_g are available. The Simulink block representing (6.7) can be developed as shown in Fig. 6.11. Do not forget to assign the initial condition in the integration block. For representation of (6.7), assign the initial condition as

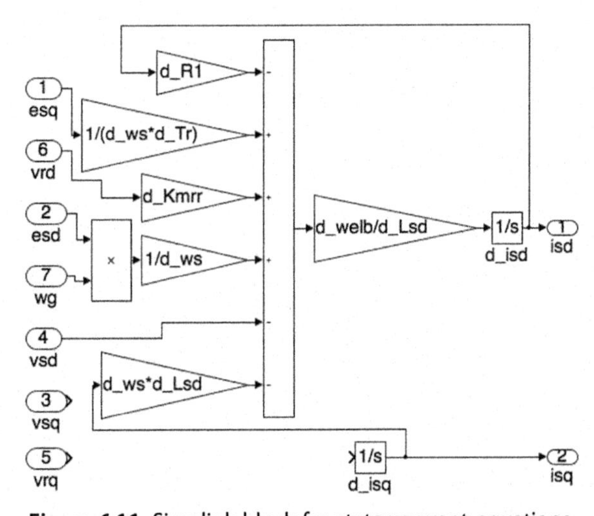

Figure 6.11 Simulink block for stator current equations.

d_isd. The representation of (6.8) is like that of Eq. (6.7) and so is omitted from Fig. 6.11 for the readers to complete. Develop a Simulink model for both equations, convert it to a subsystem and name it *isd_isq*. The *isd_isq* block should take seven inputs ($e_{sq}, e_{sd}, v_{sq}, v_{sd}, v_{rq}, v_{rd}, \omega_g$) and produce two outputs (i_{sq}, i_{sd}).

Similarly, develop subsystems *eds_eqs_idr_iqr* and *Output* subsystems as shown in Fig. 6.12. *eds_eqs_idr_iqr* represents Eqs. (6.9, 6.10, 6.17, 6.18), and the *Output* represents Eqs. (6.20), (6.21) and (6.23). Specify the initial values of state variables as d_esd, d_esq, d_isd and d_isq, in the respective integration blocks. The parameters for the model are listed in Table 6.2. Name the subsystem *Generator* as shown in Fig. 6.13.

Now assemble the turbine, generator and network blocks to form a DFIG_SMIB Simulink model as shown in Fig. 6.13. Note that the output current of DFIG consists of stator current and current through the filter. However, we have not developed the filter model yet. Hence, use constant

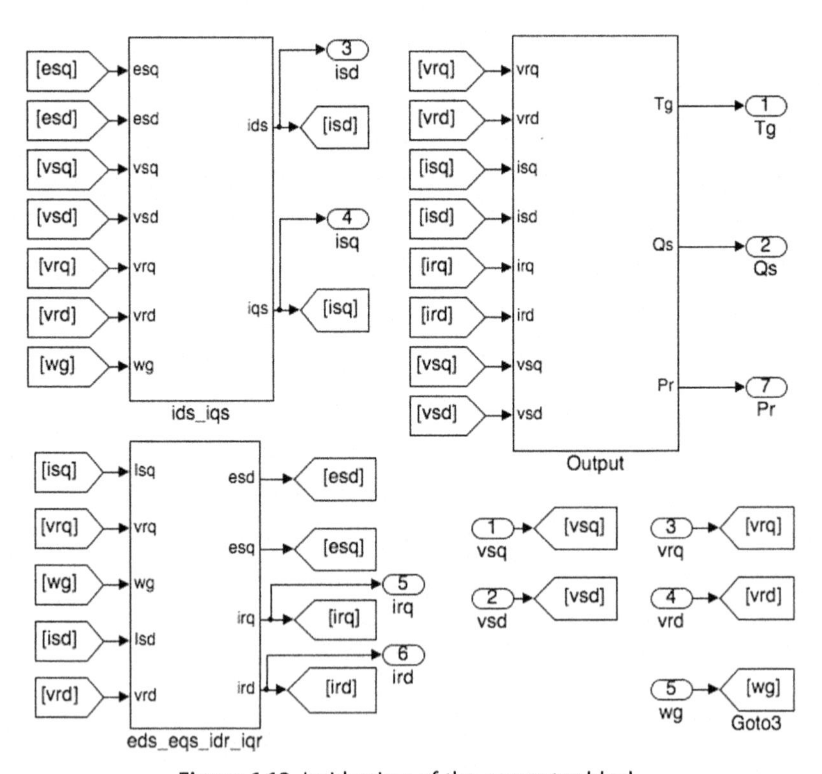

Figure 6.12 Inside view of the generator block.

blocks, d_iqg and d_idg, to represent the current output of filter. Test the model developed so far using the values in test_dfig.m. First run test_dfig.m and then run the Simulink model developed. Compare different variables in the simulation against the initial values in test_dfig. As the input remains constant, the value of state variables should not change. Two example values are shown in Fig. 6.13.

6.1.4 LCL filter

Now it is time to complete the DFIG-SMIB model by linking the rotor winding to grid, which consists of the back-to-back converter and filter. In this book, an LCL filter as shown in Fig. 6.14 is used. The filter consists of two inductors (L_i, L_g), a capacitor (C_f) and a damping resistor (R_c). The resistances in series with inductors (R_i, R_g) represent stray resistance of the inductor. Remember to use use 'd_' prefix for all the parameters and variables in the program. The design of LCL filters is discussed widely in

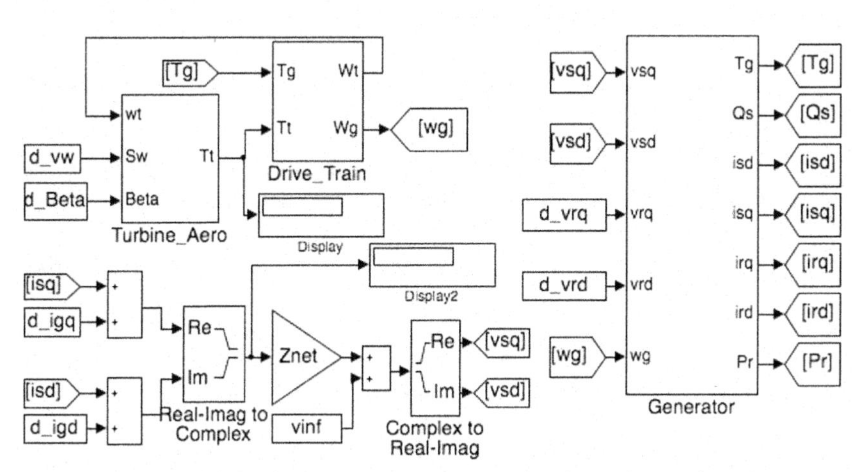

Figure 6.13 Doubly fed induction generator (DFIG)—single machine infinite bus (SMIB) Simulink model without converter and filter.

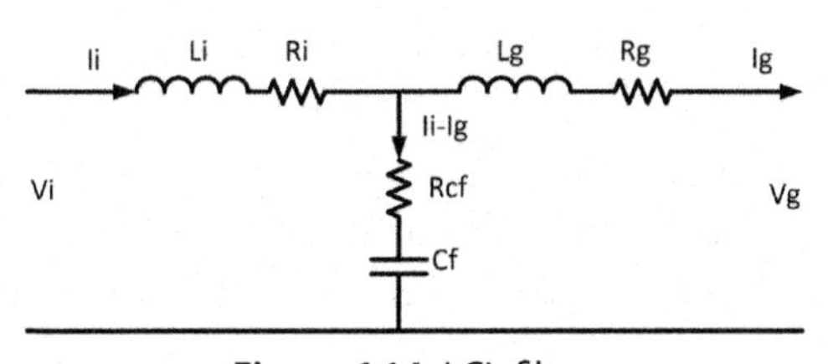

Figure 6.14 LCL filter.

Table 6.3

Parameter	**Symbol (variable name)**	**Value (p.u)**
Inverter-side inductance	L_i (d_Li)	0.1667
Inverter-side resistance	R_i (d_Ri)	0.0000
Grid-side inductance	L_g (d_Lg)	0.0033
Grid-side resistance	R_g (d_Rg)	0.0000
Filter capacitor	C_f (d_Cf)	0.0150
Damping resistance	R_c (d_Rc)	0.7333

literature (Rosyadi et al., 2012; Araujo et al., 2007), and example parameters used in this chapter are listed in Table 6.3. In the d-q synchronous reference frame, Eqs. (6.24)–(6.29) are used to represent the LCL model. In Fig. 6.1, the LCL filter connects the inverter output to Bus 3. This means the LCL filter model takes inverter voltages (v_{iq}, v_{id}) and Bus 3 voltages or stator voltage (v_{sq}, v_{sd}) as inputs and provides the current injected at Bus 3 through the filter (i_{gq}, i_{gd}) as output.

$$\frac{L_i}{\omega_b}\frac{d}{dt}i_{iq} = v_{iq} - v_{cq} - (R_i + R_c)i_{iq} + \omega L_i i_{id} + R_c i_{gq} \tag{6.24}$$

$$\frac{L_i}{\omega_b}\frac{d}{dt}i_{id} = v_{id} - v_{cd} - (R_i + R_c)I_{id} - \omega L_i I_{iq} + R_c I_{gd} \tag{6.25}$$

$$\frac{L_g}{\omega_b}\frac{d}{dt}I_{gq} = v_{cq} - v_{gq} - (R_g + R_c)I_{gq} + \omega L_g I_{gd} + R_c I_{iq} \tag{6.26}$$

$$\frac{L_g}{\omega_b}\frac{d}{dt}I_{gd} = v_{cd} - v_{gd} - (R_g + R_c)I_{gd} - \omega L_g I_{gq} + R_c I_{id} \tag{6.27}$$

$$\frac{C_f}{\omega_b}\frac{d}{dt}v_{cq} = I_{iq} - I_{gq} - \omega C_f V_{cd} \tag{6.28}$$

$$\frac{C_f}{\omega_b}\frac{d}{dt}v_{cd} = I_{id} - I_{gd} + \omega C_f V_{cq} \tag{6.29}$$

A Simulink representation of the LCL filter model is shown in Fig. 6.15, which represents (6.24)–(6.29). The Iside (inverter side), Gside (grid side) and capacitor blocks in the figure represent Eqs. 6.24–6.29, respectively. Ensure that the 'd_' prefix is used in the parameters and state variables. For example, v_{cd} must be entered as d_vcd. The figure contains two

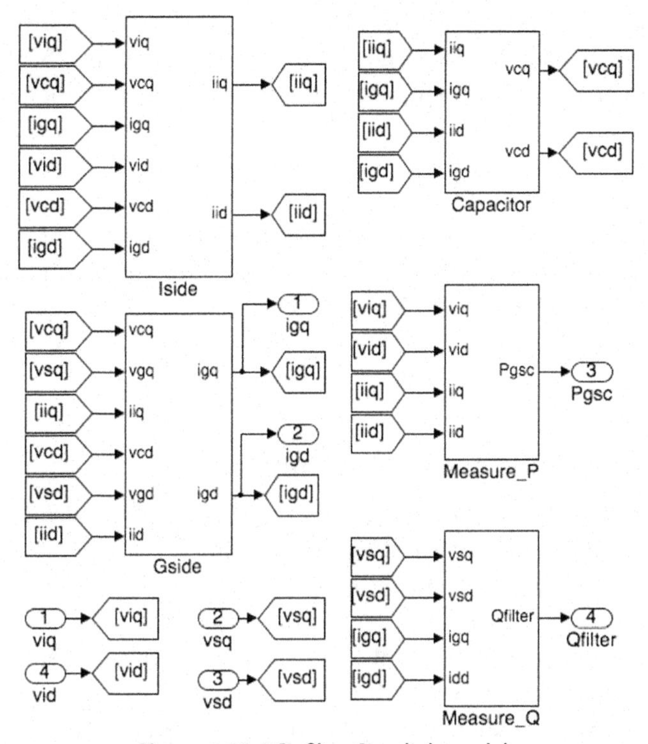

Figure 6.15 LCL filter Simulink model.

additional blocks Measure_P and Measure_Q representing (6.30) and (6.31), respectively. Eq. (6.30) represents active power output at the converter terminal, and Eq. (6.31) represents reactive power output at the filter terminal. Both outputs will be used for the GSC blocks discussed in Section 6.6.7.

$$P_{gsc} = v_{iq}i_{iq} + v_{id}i_{id} \tag{6.30}$$

$$Q_{gsc} = -v_{sq}i_{gd} + v_{sd}i_{gq} \tag{6.31}$$

6.1.5 Back-to-back capacitor

The back-to-back converter connects the rotor windings with the grid through the LCL filter. It consists of a MSC, dc link capacitor and a GSC. The switching effects of the converters are neglected. It is assumed that the converter dynamics are fast and that the converters follow the reference generated by the converter controller in real time; hence, it is sufficient to model the control alone.

Neglecting switching and conduction losses in the converter, the dynamics of the capacitor voltage can be represented using (6.32). Here, P_{msc} is the active power flowing through the MSC, P_{gsc} is the active power flowing through GSC, v_{dc} is the capacitor voltage and C_{dc} is the capacitance. In case of the DFIG, P_{msc} is equal to rotor power P_r. At steady-state, rotor power (P_r) is equal to GSC output power (P_{gsc}), so that the capacitor voltage (v_{dc}) is constant. Build a subsystem named B2BC for the capacitor with MSC power (rotor power) and GSC power as inputs to the subsystem and capacitor voltage as output of the subsystem. The block is shown in Fig. 6.21.

$$\frac{1}{C_{dc}} p v_{dc} = \frac{1}{v_{dc}} (P_{msc} - P_{gsc}) \tag{6.32}$$

6.1.6 Machine-side converter controller

There is an important distinction between the blocks so far developed and the controllers. The controller models represent a set of software code run in a microcontroller, whereas the models so far developed represent mechanical or electrical systems. The controller models receive electrical signals corresponding to voltage, current and generator rotor speed and produce switching signals for the converter. The converters use a vector control approach where independent control of active power or torque and reactive power or voltage is achieved. However, the independent control is only possible when one of the d-q axes is aligned with the stator voltage. The generator model so far developed assumes that the q-axis is aligned with infinite bus voltage. The current and voltage measurements must be aligned to a new reference frame inside the controllers. A simplified version of controllers consisting of two cascaded PI controllers will be used in this chapter. There are numerous variations of controllers proposed in the literature. Once a working simulation is developed, the readers can explore these methods and incorporate them into their simulation program.

A block diagram representation of the MSC controller is shown in Fig.6.16. It has two cascaded PI control loops in the d- and q-axis. The inner loops (MSC_IL1, MSC_IL2) are fast current controllers and outer loops (MSC_OL1, MSC_OL2) are slower torque or reactive power controllers.

Reactive power and torque reference: The reactive power reference (Qs) can be selected based on the grid requirement. In this simulation, the reactive power output reference is obtained from the power flow solution. When the DFIG is operating at the rated operating condition, the torque reference can be set to 1 p.u. However, at subrated operating condition,

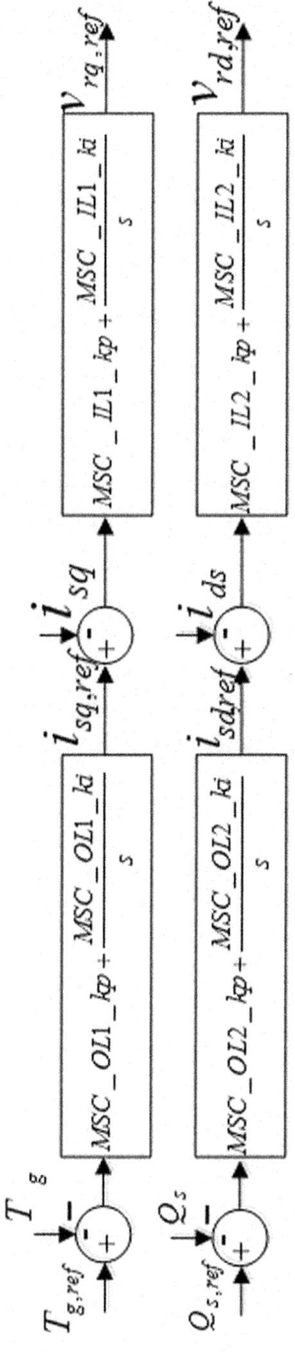

Figure 6.16 Machine-side converter (MSC) block diagram representation.

the reference torque can be set to extract maximum wind power, called maximum power point tracking (MPPT). The MPPT is achieved by setting the pitch angle to zero and controlling the turbine speed such that the tip speed ratio $\lambda = \lambda_{opt}$. Let $P_t = P_{t\,\text{rated}} T_t \omega_t$ (W), $v_w = \omega_{t\,\text{rated}} \omega_t R / \lambda_{opt}$ (m/s) and $C_p(\beta, \lambda) = C_{pmax}$ in (6.1)

$$P_{t\,\text{rated}} T_t \omega_t = 0.5 \rho \pi R^2 C_{pmax} \left(\frac{\omega_{t\,\text{rated}} \omega_t R}{\lambda_{opt}} \right)^3 \tag{6.33}$$

From (6.33), the reference value of torque is $T_t = K_{opt} \omega_t^2$ at subrated operating condition, where $K_{opt} = 0.5 \rho \pi R^5 C_{pmax} \omega_{t\,\text{rated}}^3 \Big/ \left(\lambda_{opt}^3 P_{t\,\text{rated}} \right)$. For the set of turbine parameters given in Section 6.6.2, $K_{opt} = 1$.

The Simulink representation of the MSC block is shown in Fig. 6.17. As explained before, inside the controller, the currents (i'_{rq}, i'_{rd}) and voltages (v'_{rq}, v'_{rd}) are in a new d-q reference frame where the q-axis is aligned with the stator voltage. Remember, in the model developed so far, the currents (i_{rq}, i_{rd}) and voltages (v_{rq}, v_{rd}) are in a d-q reference frame where the q-axis is aligned with the infinite bus voltage. The conversion between the two-reference frames is carried out by rotating the phasor to an angle θ, which is the angle of the stator voltage (DFIG bus voltage). The dq2dq′ and dq′2dq blocks at both the ends of Fig. 6.17 represent the conversion between the two-reference frames. For example, in the dq2dq′ block,

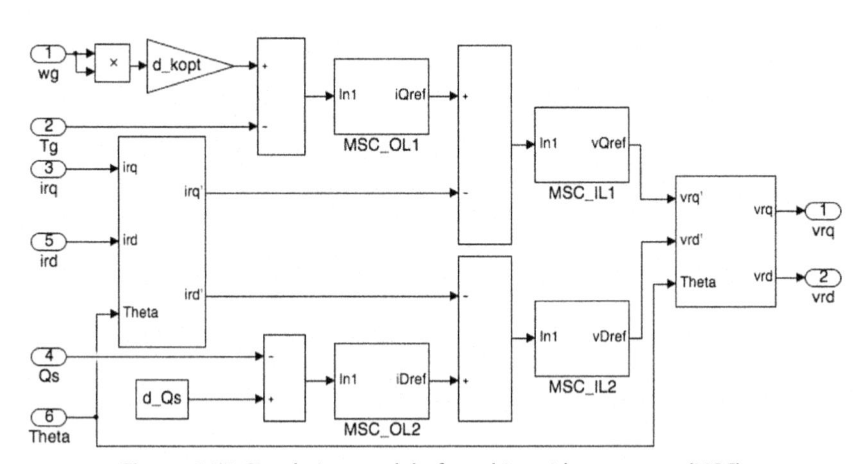

Figure 6.17 Simulation model of machine-side converter (MSC).

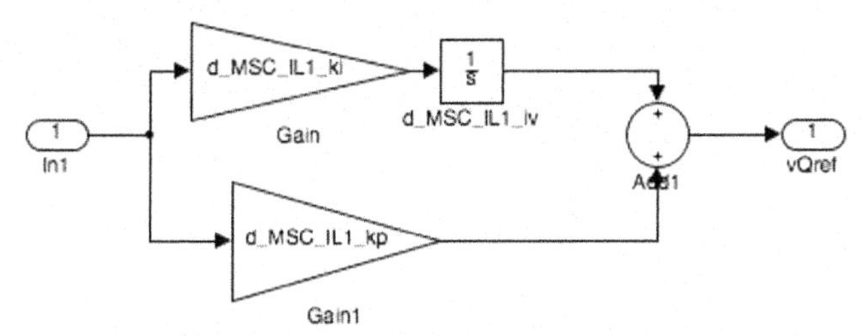

Figure 6.18 Example PI controller structure.

current is transformed using $i'_{rq} + j * i'_{rd} = (i_{rq} + j * i_{rd})e^{-j\theta}$, and in the dq'2dq block, voltage is transformed using $v_{rq} + j * v_{rd} = \left(v'_{rq} + j * v'_{rd}\right)e^{j\theta}$.

There are four PI controllers (MSC_OL1, MSC_OL2, MSC_IL1, MSC_IL2) having a common structure. Fig. 6.19 shows the structure of MSC_OL1. The gains and initial condition of the block are d_MSC_OL1_kp, d_MSC_OL1_ki and d_MSC_OL1_iv. Ensure that the correct variable names of the initial condition are specified in the integration blocks. Their values are assigned in the program given in Script 6.1. The other three PI controller blocks have same structure, but with slight change in the variable names. For example, for MSC_IL2 block, the gains and initial condition of the block are d_MSC_IL2_kp, d_MSC_ IL2_ki and d_MSC_ IL2_iv.

6.1.7 Grid-side converter controller

Fig. 6.19 shows the block diagram representation of the GSC controller. Like the RSC controller, it also has two cascaded PI controllers regulating the capacitor voltage and reactive power flow from the GSC at the WTG bus. By regulating the capacitor voltage to its reference value, the rotor power will be transferred to the grid ($P_r = P_{gsc}$). The simulation presented in this chapter will assume that no reactive power is transferred to the grid through the GSC; hence, the reactive power reference is set to zero. Fig. 6.20 shows the Simulink representation of the GSC controller, which can be built like the RSC controller. The structure of the PI controllers is like the one in Fig. 6.18. The gains and initial condition names must be changed. For example, in GSC_OL1 blocks, the values are d_GSC_OL1_kp, d_GSC_OL1_ki and d_GSC_OL1_iv. As explained earlier, the voltage and current phasors are aligned to the new d-q reference

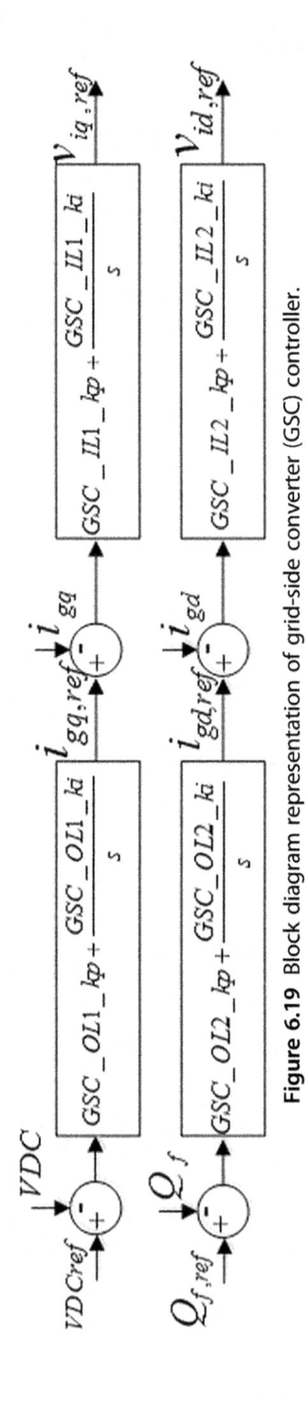

Figure 6.19 Block diagram representation of grid-side converter (GSC) controller.

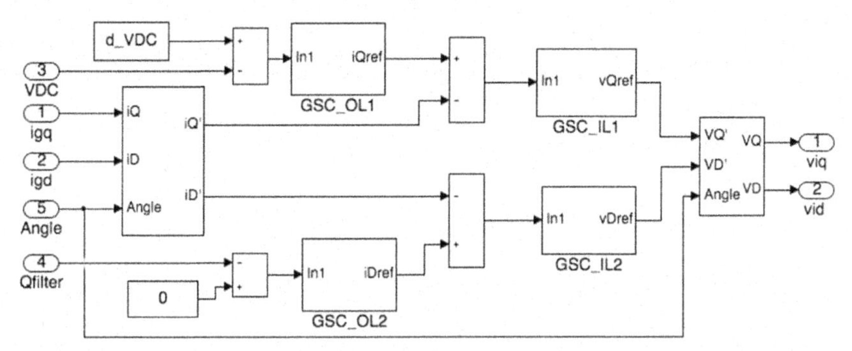

Figure 6.20 Simulation representation of grid-side converter (GSC) controller.

frame inside the controller to satisfy the independent active and reactive power control requirement.

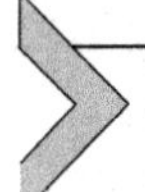

6.3 Single machine infinite bus model integration and testing

Now all the Simulink blocks required for the SMIB simulation are ready, and it is time to integrate the model and test it. Connect the blocks so far developed as shown in Fig. 6.21 and save the model as smib_dfig_sim. Run test_dfig.m and run the simulation. Validate the output against Fig. 6.21. Now linearize the model using the *linmod* command and calculate the eigenvalues using the *eig* command as described in Chapter 4. Table 6.4 shows the eigenvalues of the SMIB system for comparison. If you are getting the same eigenvalues, the model developed is correct. If it is not, try to debug it using the procedure explained in Section 3.6.

6.3.1 Dynamic simulation

Let us run a dynamic simulation. Make two changes in the model. (1) Replace d_vw constant block with a step block. Open the block parameter dialogue and make changes: step time = 1, initial value = d_vw and final value = d_vw-2. The changes mean at time $t = 1$ s, the wind speed will reduce by 2 m/s from its initial value. (2) Replace vinf block by a step input block and make changes in its block parameter dialogue. Step time = 10 s, initial value = vinf and final value = vinf+0.02. Now change the simulation stop time to 20 s and run the model. Some of the example outputs are shown in Fig. 6.22. Are you getting same outputs? Congratulations. Your model is ready now.

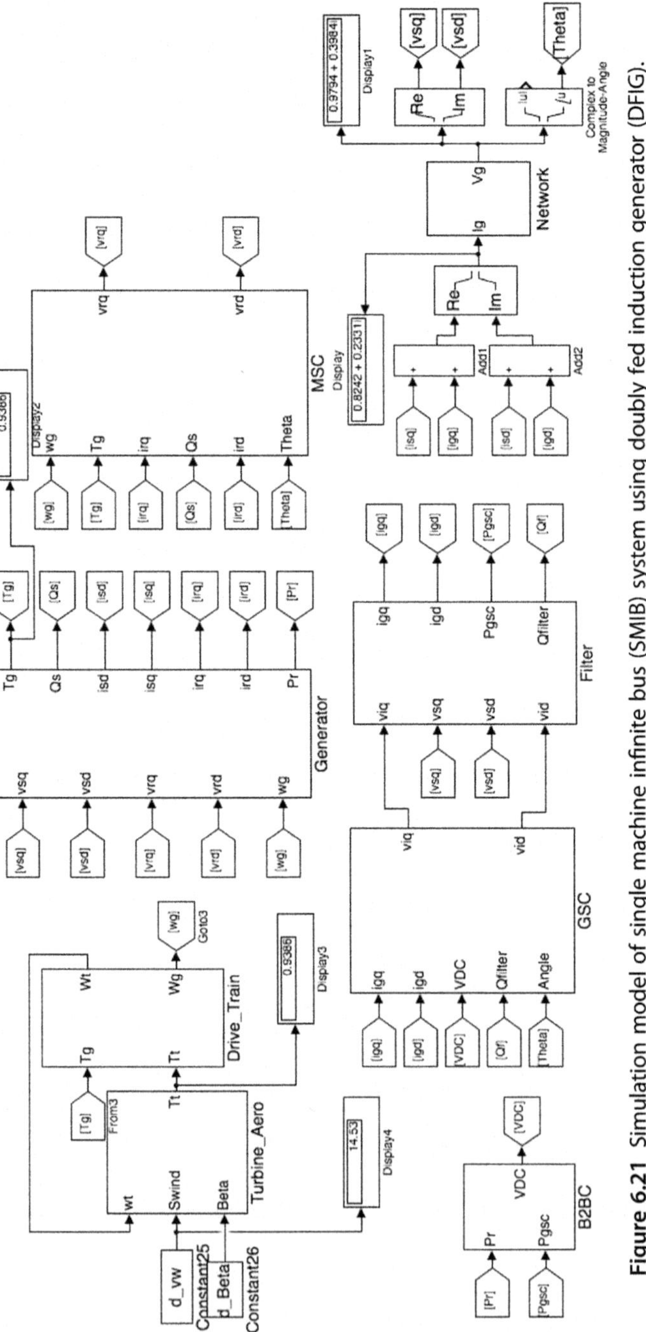

Figure 6.21 Simulation model of single machine infinite bus (SMIB) system using doubly fed induction generator (DFIG).

Table 6.4 Eigenvalues of the single machine infinite bus (SMIB)—doubly fed induction generator (DFIG) system.

Eigenvalues	Frequency (Hz)	Damping ratio (%)
$-58581 \pm 62892i$	10010	68.15
$-16564 \pm 17901i$	2849.1	67.92
$-674.3 \pm 1979.9i$	315.11	32.23
$-322.3 \pm 645.5i$	102.73	44.67
$-211.4 \pm 335.4i$	53.37	53.32
$-79.40 \pm 97.15i$	15.46	63.28
-62.12	0	100
$-37.76 \pm 74.71i$	11.89	45.10
$-4.15 \pm 16.91i$	2.69	23.87
-12.91	0	100
-11.54	0	100
$-2.94 \pm 11.01i$	1.75	25.85
-0.33	0	100

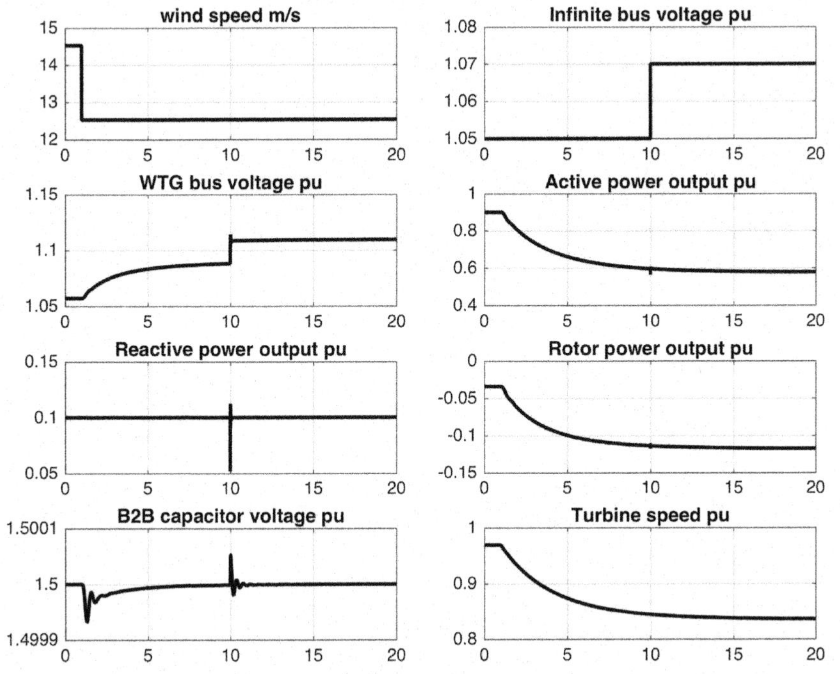

Figure 6.22 Dynamic simulation results of doubly fed induction generator (DFIG)—single machine infinite bus (SMIB) system.

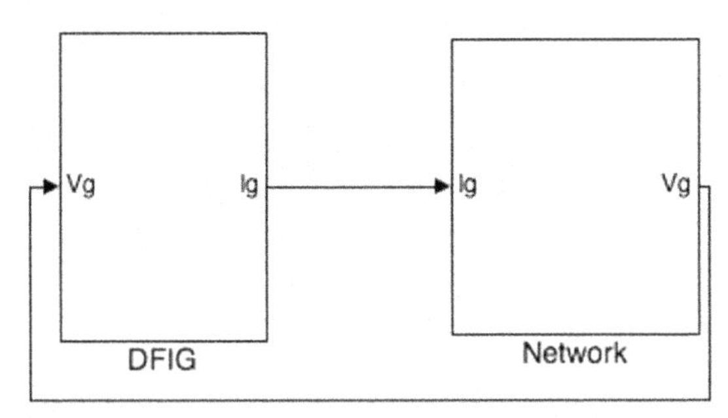

Figure 6.23 Complete single machine infinite bus (SMIB)—doubly fed induction generator (DFIG) simulation model.

So far we have used the initial values in Script 6.1 for the Simulink model. It corresponds to a power output of 0.8 + 0.1j pu. This is not very helpful, as we may like to run the model at different operating conditions, which requires different sets of initial values. Section 6.4 discusses the steps required to find initial values of state variables at any operating condition.

Before we move to initialization, let us create a subsystem named DFIG as shown in Fig. 6.23. The network block takes complex value of DFIG output current and outputs DFIG bus voltage. The DFIG block can be readily used in the wind farm simulation model developed in Section 6.9.

6.4 Initialization of SMIB-DFIG system

The initialization program finds the initial value of all state variables in the system, which is used in the integration blocks. The program for the DFIG initialization is given in Scripts 6.2–6.4. The program can be divided into the following simple steps.

Step 1: Run the power flow program to determine the voltage and power injection at the WTG bus.

```
Dmachs = [3];      % Bus where PMSG is connected
Omega = 2*pi*50;
if size(Pmachs,1)
      find_dfig_state_initial_conditions
end
```

Script 6.3 Program to call doubly fed induction generator (DFIG) initialization.

```
%STEP 2 BEGINS: Gather machine parameter for DFIG
d_welb = 2*pi*50; % electrical base speed rad/sec
d_ws = 1.0;       % Synchronous speed in pu

% DFIG parameters
%data_DF= [mach# bus# d_Lm  d_Rs   d_Rr    d_Lss d_Lrr   d_Kopt d_ktg d_ctg  d_Ht d_Hg BaseMVA]
data_DF = [1      5    4     0.005 0.0055 4.04  4.0602 1       0.3   0.01   4     0.4  5];

% DFIG-Filter parameters
%FLTR = [d_Ri, d_Rg, d_Rc,  d_Li,  d_Lg,  d_Cf];
d_FLTR =  [0.000, 0.000, 0.7333, 0.1667, 0.0033, 0.0150];
d_Cdc = 2; % Converter capacitor

%DFIG-MSC Controller Parameters
d_MSC_IL1_kp = -0.23;  d_MSC_IL1_ki = -3;
d_MSC_IL2_kp = -0.23;  d_MSC_IL2_ki = -3;
d_MSC_OL1_kp = 0;      d_MSC_OL1_ki = -60;
d_MSC_OL2_kp = 0;      d_MSC_OL2_ki = 90;

% DFIG-GSC Controller Parameters
d_GSC_IL1_kp = 0.3; d_GSC_IL1_ki = 200;
d_GSC_IL2_kp = 0.3; d_GSC_IL2_ki = 200;
d_GSC_OL1_kp = -22; d_GSC_OL1_ki = -870;
d_GSC_OL2_kp = 0;   d_GSC_OL2_ki = -60;

% Read filter parameters from FLTR vector
d_Ri= 0.0;
d_Rg= 0.0;
d_Rc= d_FLTR(3);
d_Li= d_FLTR(4);
d_Lg= d_FLTR(5);
d_Cf = d_FLTR(6);

% Read machine data from data_DF
d_Lm = data_DF(3);
d_Xm = d_Lm;
d_Rs = data_DF(4);          % stator resistance
d_Rr = data_DF(5);          % rotor resistance
d_Lss = data_DF(6);         % stator self inductance
d_Lrr = data_DF(7);         % rotor inductance
d_kopt = data_DF(8);
d_ktg = data_DF(9);
d_ctg = data_DF(10);
d_Ht = data_DF(11);
d_Hg = data_DF(12);
d_base = data_DF(13);

d_bl = 40.05;     % blade length
d_wtrated = 3.0337;  % rated turbine speed
rho = 1.225;  % air density

% Calcuating derived variable using equation () - ()
d_Ls_d = d_Lss - (d_Lm^2/d_Lrr);
d_Kmrr = d_Lm/d_Lrr;
d_R2 = d_Kmrr^2*d_Rr;
d_R1 = d_Rs + d_R2;
d_Tr = d_Lrr/d_Rr;
```

Script 6.4 Program for find_dfig_state_initial_conditions.m.

```
    % STEP 2 ENDS

    %STEP 3 BEGINS: Initialization of State variables for generator,
%Converter and Filter
    Vdfig = bus_sln(Dmachs,2).*exp(1i*bus_sln(Dmachs,3)*pi/180);
    Pdfig = bus_sln(Dmachs, 4);
    Qdfig = bus_sln(Dmachs, 5);
    d_vsq = real(Vdfig);        d_vsd = imag(Vdfig);
    d_Theta = angle(Vdfig);

xdfig  = zeros(size(Dmachs,1),15);
    for d_index = 1:size(Dmachs,1)
        xdfig0 = ones(1,15);
        if Pdfig(d_index)<1     % If below rated speed

% The function fsolve is used to solve set of SSCs described in the
%function init_dfig_mpt. The output is stored in xdfig variable
            xdfig(d_index,:) = fsolve(@(x)...
init_dfig_mpt(x,Vdfig(d_index),Pdfig(d_index),...
Qdfig(d_index),data_DF,d_FLTR),...
            xdfig0,optimset('TolFun',1e-16,'TolX',1e-16));
        else
% The function fsolve is used to solve set of SSCs described in the
%function init_dfig_cpt. The output is stored in xdfig variable

            xdfig(d_index,:) = fsolve(@(x)...
init_dfig_cpt(x,Vdfig(d_index),Pdfig(d_index),...
Qdfig(d_index),data_DF,d_FLTR),...
            xdfig0,optimset('TolFun',1e-16,'TolX',1e-16));
        end
    end

    d_isq =  xdfig(:,1);      d_isd = xdfig(:,2);
    d_irq =  xdfig(:,3);      d_ird = xdfig(:,4);
    d_vrq =  xdfig(:,5);      d_vrd = xdfig(:,6);
    d_iiq  = xdfig(:,7);     d_iid  = xdfig(:,8);
    d_igq  = xdfig(:,9);     d_igd  = xdfig(:,10);
    d_viq  = xdfig(:,11);    d_vid  = xdfig(:,12);
    d_vcq  = xdfig(:,13);    d_vcd  = xdfig(:,14);
    d_wg   = xdfig(:,15);

% d_esq and d_esd are calculated using (6.12) and (6.13)
    d_esq =  d_Kmrr.*d_ws.*(d_Lrr.*d_ird + d_Lm.*d_isd);
    d_esd = -d_Kmrr.*d_ws.*(d_Lrr.*d_irq + d_Lm.*d_isq);

    %******** STEP 4 BEGINS: Initialization of converter controllers
%and turbine  *******
    % Parameters for RSC controller model
    d_vr_dash =(d_vrq+1i*d_vrd).*exp(-1i*d_Theta);
    d_ir_dash =(d_irq+1i*d_ird).*exp(-1i*d_Theta);
    d_MSC_IL1_iv = real(d_vr_dash);
    d_MSC_IL2_iv = imag(d_vr_dash);
    d_MSC_OL1_iv = real(d_ir_dash);
    d_MSC_OL2_iv = imag(d_ir_dash);

    d_Qs = Qdfig;    % Reactive power reference RSC controller


    % B2B capacitor
    d_VDC = 1.5;
```

Script 6.4 ***(continued)*.**

```
        % Parameters for GSC controller model
        d_vi_dash = (d_viq+1i*d_vid).*exp(-1i*d_Theta);
        d_ii_dash =  (d_igq+1i*d_igd).*exp(-1i*d_Theta);

        d_GSC_IL1_iv = real(d_vi_dash);
        d_GSC_IL2_iv = imag(d_vi_dash);
        d_GSC_OL1_iv = real(d_ii_dash);
        d_GSC_OL2_iv = imag(d_ii_dash);

        d_Qfilter = 0;  % Reactive power reference GSC controller

        % Turbine initialisation
        d_Tg = d_esq.*d_isq+d_esd.*d_isd;
        d_Ts = d_Tg;
        d_wt = d_wg;
        d_Pt = d_Ts.*d_wt;

        for d_index = 1:size(Dmachs,1)

            if d_wg(d_index)>=1    % Above rated wind speed operation
                d_Sw(d_index) = 15;
                d_Lambda = 3.0337*d_wg(d_index).*d_bl/d_Sw(d_index);
    Cp_req = (d_base*d_Pt(d_index)*1e6)/(0.5*1.225*pi*d_bl^2*d_Sw(d_index)^3);
                d_Beta(d_index) = find_beta(Cp_req, d_Lambda);
            else                        % Below rated wind speed
                d_Lambda = 8.1;
                d_Beta = 0;
d_Sw(d_index,1) = (d_Pt(d_index)*d_base*1e6/(0.5*1.225*pi*d_bl^2*0.48))^(1/3);

            end
        end
```

Script 6.4 (*continued*).

Script 6.2 contains the program for this step. The system data (bus_sln, line_sln) is obtained using power flow code in Chapter 2. The *form_Ymatrix* and *power_flow* functions explained in Chapter 2 must be saved in the current working directory. Create a new file dfig_smib_main.m by combining Scripts 6.1 and 6.3.

Create a new .m file named find_dfig_state_initial_conditions.m and copy the program from Script 6.4. The script is divided into the following steps as discussed below.

Step 2: Gather machine parameters. The generator and turbine parameters are specified as data_DF vector. Each column of data_DF represents machine id, bus number, L_m, R_s, R_r, L_s, L_r, k_{opt}, k_{tg}, c_{tg}, H_t, H_g and base MVA at which the data are specified. Filter data are bundled to vector FLTR where the elements represent R_i, R_g, R_c, L_i, L_g and C_f.

Step 3: Initialization of state variables for generator, converter and filter.

The differential equations representing the generator and filter are presented in the previous section. Steady-state equations of the model can be obtained by setting the differential term $\left(\frac{d}{dt}\right)$ of the equation to zero. For easy reference in the Matlab script, let us call these equations steady-state conditions (SSCs). We will define 15 of them for a subrated operating condition.

SSC 1–4: The first four conditions represent the steady-state voltage equations of the DFIG given by (6.34)–(6.37).

$$v_{qs} = -R_s i_{qs} + \omega_s(L_{ss} i_{ds} + L_m i_{dr}) \tag{6.34}$$

$$v_{ds} = -R_s i_{ds} - \omega_s(L_{ss} i_{qs} + L_m i_{qr}) \tag{6.35}$$

$$v_{qr} = -R_r i_{qr} + s\omega_s(L_{rr} i_{dr} + L_m i_{ds}) \tag{6.36}$$

$$v_{dr} = -R_r i_{dr} - s\omega_s(L_{rr} i_{qr} + L_m i_{qs}) \tag{6.37}$$

where $\omega_r = (1 - s)\omega_s$.

SSC 5: At steady state, the electrical torque (6.24) produced by the machine under subrated operating condition is equal to $K_{opt}\omega_t^2$. At rated operating condition, this is equal to 1.

SSC6: Sum of active power output through stator and rotor windings $(P_s + P_r)$ in (6.20) and (6.21) is equal to active power injection at the WTG bus, which is obtained from the power flow solution.

SSC 7: The reactive power generation from stator in (6.22) is equal to reactive power injection at the WTG bus, which is obtained from the power flow solution. This assumes that no reactive power is supplied to grid through the filter path.

SSC 8: The reactive power injection at the WTG bus through the filter is zero, i.e., $-v_{sq}i_{gd} + v_{sd}i_{gq} = 0$.

SSC 9: Assuming no converter losses, rotor output power $(v_{rq}i_{rq} + v_{rd}i_{rd})$ is equal to GSC output power $(v_{iq}i_{iq} + v_{id}i_{id})$.

SSC 10–15: The last six equations represent the steady-state equations of the LCL filter, which are obtained by setting the left-hand side of Eqs. (6.24) –(6.29) to zero.

The SSCs 1–15 can be represented as $F = Ax$ and solved for x using the fsolve function in Matlab. The Script 6.5 represents a function representing the $F = Ax$ equation, which returns vector F. Copy the function and save it as init_dfig_mpt.m. Once init_dfig_mpt function is ready, the fsolve function in Matlab can be used to solve the 15 equations representing the 15 SSCs. The solution of the SSCs is stored in the variable xdfig. e'_{sq} and e'_{sd} are calculated using (6.12) and (6.13)

```
function F = init_dfig_mpt(x,Vdfig,Pg,Qg, data_DF, FLTR)
% This function is used to initialize DFIG for maximum power point
%tracking
% region. The prefix 'd_' is not used in the variables inside the
%function
%for ease of reading.The variables defined inside function are not
%visible
%outside it.

% Get voltage, and power from Vdfig
vqs = real(Vdfig);
vds = imag(Vdfig);
ws = 1.0;      % synchronous speed

% Get machine parameters from data_DFIG matrix
Lm = data_DF(:,3);
Rs = data_DF(:,4); % stator resistance
Rr = data_DF(:,5); % rotor resistance
Lss = data_DF(:,6); % stator self inductance
Lrr = data_DF(:,7); % rotor inductance
kopt = data_DF(8);

% Get filter parameters from FLTR matrix
R1f = FLTR(1);          R2f = FLTR(2);          Rcf = FLTR(3);
L1f = FLTR(4);          L2f = FLTR(5);          Cf = FLTR(6);

% x is the solution vector. The commented section below shows index
%of
% various parameters
% x(1) = iqs, x(2) = ids, x(3) = iqr, x(4) = idr
% x(5) = vqr, x(6) = vdr
% x(7) = i1q, x(8) = i1d, x(9) = i2q, x(10) = i2d
% x(11) = v1q, x(12) = v1d, x(13) = vcq, x(14) = vcd  x(15) = wr

% set of Ax-b = 0 equations. Refer Section xx for details.
F = [-Rs*x(1)+ws*(Lss*x(2)+Lm*x(4))-vqs;    % SSC 1
    -Rs*x(2)-ws*(Lss*x(1)+Lm*x(3))-vds;     % SSC 2
    -Rr*x(3)+(ws-x(15))*(Lrr*x(4)+Lm*x(2))-x(5);  % SSC 3
    -Rr*x(4)-(ws-x(15))*(Lrr*x(3)+Lm*x(1))-x(6);  % SSC 4
    Lm*(x(1)*x(4)-x(2)*x(3))-kopt*x(15)^2;  % SSC 5
    vqs*x(1)+vds*x(2)+vqs*x(9)+vds*x(10)-Pg;  % SSC 6
    -vqs*x(2)+vds*x(1)-Qg;                 % SSC 7
    -vqs*x(10)+vds*x(9);                 % SSC 8
    x(5)*x(3)+x(6)*x(4)-x(11)*x(7)-x(12)*x(8);  % SSC 9
    x(7)*(R1f+Rcf) - x(8)*L1f - x(9)*Rcf + x(13) - x(11); % SSC 10
    x(8)*(R1f+Rcf) + x(7)*L1f - x(10)*Rcf + x(14) - x(12); % SSC 11
    -x(7)*Rcf + x(9)*(Rcf+R2f) - x(10)*L2f + vqs - x(13); % SSC 12
    -x(8)*Rcf + x(10)*(Rcf+R2f) + x(9)*L2f + vds - x(14); % SSC 13
    x(7) - x(9) + Cf*x(14); % SSC 14
    x(8)-x(10)-Cf*x(13)];   % SSC 15
```

Script 6.5 Function used to initialize doubly fed induction generator (DFIG) under subrated operating condition.

Step 4: Initialization of converter controllers and turbine. In this step, the initial values of converter controllers and turbine are calculated. At subrated condition, the wind speed is calculated using Eq. (6.1) by setting $\beta = 0$ and $\lambda = \lambda_{opt}$. We will discuss the method to obtain turbine initial condition for rated operating condition in the next section.

The programs required for the initial condition calculation can be summarized as follow:

(1). Make form_Ymatrix.m and power_flow.m as discussed in Chapter 2.
(2). Make dfig_smib_main.m by combining Scripts 6.1 and 6.3.
(3). Make find_dfig_state_initial_conditions.m from Script 6.4.
(4). Make init_dfig_mpt.m using the Script 6.5.

Ensure that all the codes are saved in one folder. To simulate different network or DFIG operating conditions, make changes in the bus or line matrices. For example, to change the output of DFIG, change bus (3,4) element. But do not make it unity, as we have not discussed the rated operating condition scenario.

6.5 Further modifications in DFIG-WTG model

So far, we have cut many corners to quickly build and run a DFIG simulation. Let us discuss couple of modifications and additions that can improve the model.

Simulating rated operating condition: The program init_dfig_mpt.m can initialize the DFIG to subrated operating condition. The program should be modified to work for rated operating condition. Modify SSC5 such that the torque is unity at rated operating condition and save it as init_dfig_cpt.m.

```
Lm*(x(1)*x(4)-x(2)*x(3))-1;   % SSC 5 for rated operating condition
```

Secondly, we need to find the wind speed and pitch angle at rated operating condition. At rated operating conditions, wind speed can take values between 15 and 25 m/s for the given turbine parameters. Once wind speed is selected, find λ using the turbine (generator) speed obtained from the generator initialization. Now write a function find_beta(Cp_req, d_Lambda) that will return β using Eq. (6.2).

MSC and GSC controllers: Adding feed forward path in the MSC and GSC controllers can enhance their performance. The structure of the controllers is available in many literatures (Rosyadi et al., 2014; Wu et al., 2007). What would be the initial condition of the MSC and GSC controller state variable when feed forward path is introduced?

Pitch angle controller: Add a pitch angle controller to simulate the rated operating condition. A block diagram for the pitch angle controller is shown in Fig. 6.24.

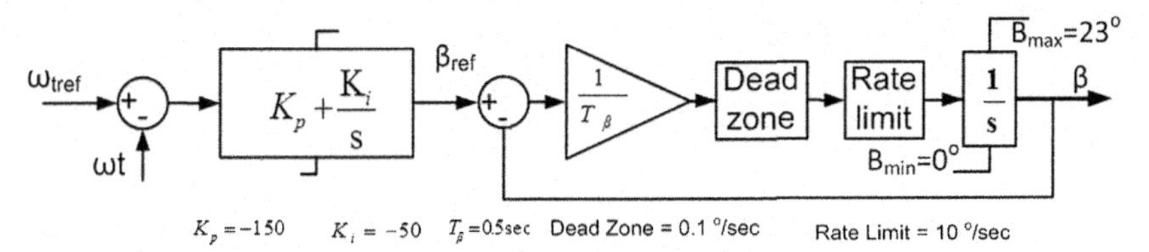

Figure 6.24 Block diagram representation of pitch angle controller.

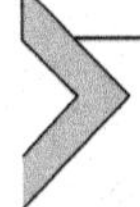

6.6 Permanent magnet synchronous generator modelling

This section presents the steps to build a Simulink program for a PMSG-based WTG connected to an infinite bus as shown in Fig. 6.25. It is assumed that the readers have already built a DFIG-SMIB model as discussed in the previous section. We will reuse many of the blocks from the DFIG model.

The PMSG uses permanent magnets to generate rotor field. They may come with or without the gearbox. A gearless direct drive configuration is widely used for offshore applications. Increasing the number of poles of generator, the operating speed can be matched with turbine speed to avoid gearbox in the design. Note that the generator frequency and speed are independent of the grid frequency. The generator is fed from a back-to-back converter, which is designed to carry rated output power. It also contains a filter to remove switching frequency components in the current.

Like DFIG-SMIB model, the PMSG-SMIB model can be divided into seven components or sections: turbine, generator, MSC, B2BC, GSC, filter and network. Let us make the Simulink model of the system first.

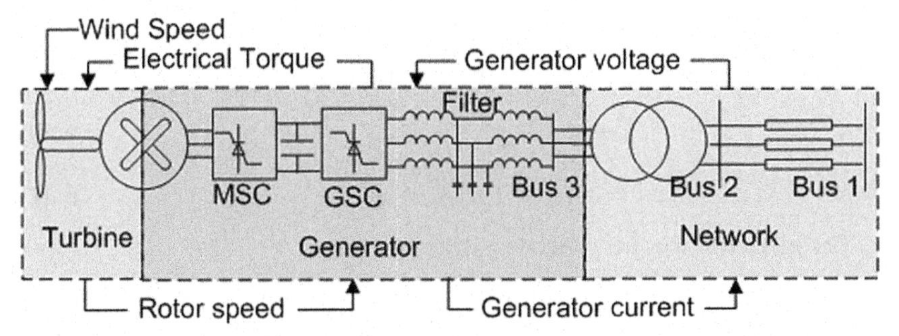

Figure 6.25 Schematic diagram of a single machine infinite bus (SMIB) system using permanent magnet synchronous generator (PMSG)–type wind turbine generator.

6.6.1 Turbine model

For continuity from DFIG model, we will assume that the turbine parameters are same for both DFIG-SMIB and PMSG-SMIB systems. This will allow us to reuse the wind turbine aerodynamic model discussed in Section 6.2.2.2. However, because of the gearless configuration, a one mass model will be used for the drive train model.

Let us start building the model from the turbine side of the system. For the aerodynamic model of turbine, use the blocks developed in Section 6.2a by changing the prefix of all parameters from 'd_' to 'p_'. A single-mass model of the drive train can be developed using the following equation:

$$\frac{d\omega_t}{dt} = \frac{1}{2H}(T_m - T_e)$$

where ω_t, H, T_m and T_e are turbine speed, inertia constant, mechanical torque and electrical torque. Note that the state variable in the one-mass model is p_wt. Now, build a Simulink block—representing turbine that takes wind speed and electrical torque as input and turbine speed as output.

6.6.2 Permanent magnet synchronous generator model

Eqs. (6.38–6.43) represent widely cited electrical model of PMSG in synchronous reference frame, where v, ψ, ω_{re}, R_a, L, i, ϕ_{pm} represent voltage, flux, speed in electrical radians/second, stator resistance, inductance, current and permanent magnet flux, respectively (Kim et al., 2010; Li et al., 2010).

$$v_d = p\psi_d - \psi_q\omega_{re} - R_a i_d \tag{6.38}$$

$$v_q = p\psi_q + \psi_d\omega_{re} - R_a i_q \tag{6.39}$$

where $\psi_d = -L_d i_d + \phi_{pm}$ and $\psi_q = -L_q i_q$.

Equation (6.38) and (6.39) can be rewritten as

$$L_d p i_d = -v_d + L_q i_q \omega_{re} - R_a i_d \tag{6.40}$$

$$L_q p i_q = -v_q - L_d i_d \omega_{re} + \phi_{pm}\omega_{re} - R_a i_q \tag{6.41}$$

The electrical torque generated is given by

$$T_e = -L_d i_d i_q + \phi_{pm} i_q + L_q i_q i_d \tag{6.42}$$

The active power output at the PMSG terminal is given by

$$P_{pmsg} = v_d i_d + v_q i_q \tag{6.43}$$

Now, build a generator model subsystem using Eqs. (6.40–6.43). The inputs to this model will be the voltages (v_q, v_d) obtained from MSC blocks and turbine speed (ω_t) obtained from turbine block. The outputs of generator block are i_d, i_q, T_e and P_{pmsg}.

6.6.3 Machine-side converter controller

Like the controllers in the DFIG model, the MSC controller in PMSG uses a decoupled control strategy, where q-axis regulates torque and d-axis regulates reactive power. The d-q axis is selected such that $v_q = |V|$ and $v_d = 0$. Let us assume the reactive power at the MSC terminal ($-v_q i_d + v_d i_q$) is zero. This is true when the d-axis current (i_d) is zero. Hence, the d-axis current reference to zero. The torque reference is obtained using the MPPT discussed in the DFIG-SMIB modelling. For $i_d = 0$, from (6.42), the electrical torque $T_e = k_{opt}\omega_t^2 = \phi_{pm} i_q$. Hence, reference value of $i_q = K_{opt}\omega_t^2 / \phi_{pm}$. The Simulink model representation of the MSC controller is shown in Fig. 6.26. It has two PI controllers, MSC_IL1 and MSC_IL2, like the DFIG-SMIB model. The gains and initial state variable must be specified like the DFIG-SMIB model with 'p_' suffix. For example, MSC_IL1 has variables p_MSC_IL1_kp, p_MSC_IL1_ki and p_MSC_IL1_iv.

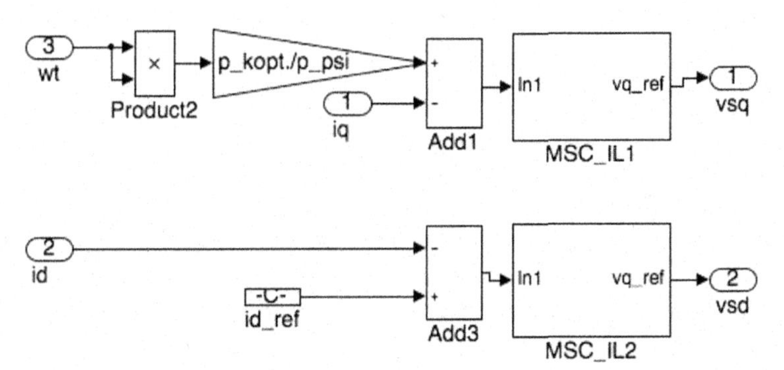

Figure 6.26 Simulink representation of permanent magnet synchronous generator (PMSG) machine-side converter (MSC).

Table 6.5 Parameters of LCL filter in permanent magnet synchronous generator (PMSG)—type wind turbine generator (WTG).

Parameter	Symbol (variable name)	Value (p.u.)
Inverter-side inductance	L_i	0.1667
Inverter-side resistance	R_i	0.0
Grid-side inductance	L_g	0.0033
Grid-side resistance	R_g	0.0
Filter capacitor	C_f	0.0150
Damping resistance	R_c	0.7333

6.6.4 Back-to-back capacitor, GSC controller, LCL filter and network

The model of B2BC, GSC and filter developed for DFIG can be directly copied to use in a PMSG simulation as the working principles of these components are same. Only (in fact the most important) change required in the Simulink blocks is in the prefix of name of parameters and state variables, which must be changed from 'd_' to 'p_'. In the DFIG, the filter connects GSC at the rotor circuit with the generator bus, which carry maximum of 30% of the rated power. However, in PMSG, it is designed to carry full rated output. For this reason, new set of parameters listed in Table 6.5 are used for the filter in PMSG.

Network model: In the DFIG model, the sum of stator current and filter current is fed to the network model. However, in PMSG_SMIB, the filter current alone is fed to the network model. The network block of DFIG-SMIB system can be copied to be adapted and used in the PMSG-SMIB system.

So far, we have discussed all the blocks required to build a PMSG-SIMB system simulation program. Integrate various blocks as shown in Fig. 6.27. Let us write a program to find initial values of the state variables. We will test the Simulink program later.

6.7 Initialization of PMSG-SMIB system

The program for initializing the system is given in Scripts 6.6 and 6.7. Create a new file pmsg_smib_main.m by combining Scripts 6.1 and 6.6.

Create a new .m file named find_pmsg_state_initial_conditions.m and copy the program in Script 6.7. The comments in the program explain the calculation steps.

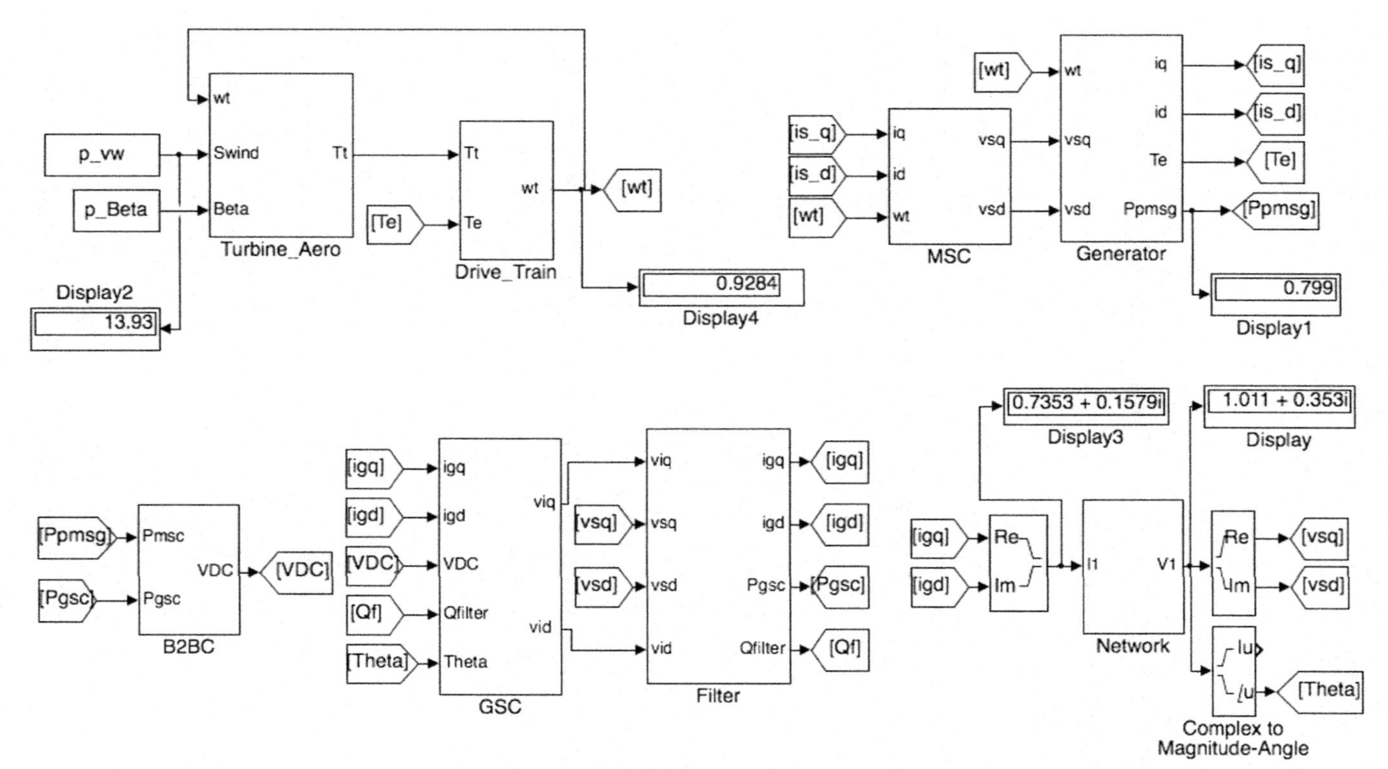

Figure 6.27 Simulink model of permanent magnet synchronous generator (PMSG)—single machine infinite bus (SMIB) system.

```
Pmachs = [3];      % Bus where PMSG is connected
Omega = 2*pi*50;
if size(Pmachs,1)
    find_pmsg_state_initial_conditions
end
```

Script 6.6 Program to call permanent magnet synchronous generator (PMSG) initialization.

6.8 Modal analysis and dynamic simulation results

The eigenvalues of the test system for the data given in Scripts 6.1 and 6.7 are listed in Table 6.6. Find eigenvalues of the Simulink model and compare with the values in the table. In case you are not getting the correct eigenvlaues, debug the Simulink model using the method discussed in Section 3.6. Let us run a dynamic simulation. Apply changes mentioned in Section 6.3. Fig. 6.28 shows dynamic simulation results by applying a step change in wind speed at time = 1 s and step change in infinite bus voltage at time = 10 s.

Finally, make a PMSG block like DFIG block in Fig. 6.23. To do this, select all blocks in Fig. 6.27 except the network block, click the right mouse button and select the 'create subsystem from selection' option. Change the name of new subsystem to PMSG. We will use this block for building a wind farm simulation program in the next section.

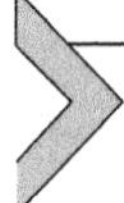

6.9 Simulation of wind farm having DFIG- and PMSG-type WTGs

So far, we have discussed modelling and simulation of a DFIG- and a PMSG-type WTG connected to an infinite bus. However, an actual wind farm contains tens or hundreds of WTGs connected to a medium voltage network called collector system. Depending on study requirement, we may like to model all the WTGs together or part of the system. This section will discuss how to use the model so far developed to build a simulation model for a big practical wind farm.

Let us take an example wind farm as shown in Fig. 6.29. It contains six WTGs, three of them (at buses 1, 2 and 3) are DFIG type and other (at buses 4, 5 and 6) three are PMSG type. There are 13 buses in the system. Consider Bus 13 as infinite bus. The program to initialize the model is given in Script 6.8. In the initial part of the program, bus and line matrices are specified, and load flow solution and admittance matrix are obtained. In the simulation programs discussed so far, one WTG is connected to

```
    %BEGINS: Parameter for PMSG
    p_ra = 0.0025; p_Lq = 0.7; p_Ld = 0.7;
    p_H = 2;
    p_psi = 1.222;
    p_kopt = 1;
    p_bl = 40.05;
    p_lambda_opt = 8.1;
    p_wt_rated = 3.0337;
    p_base = 5;

    % Filter parameters
    p_FLTR =  [0.000, 0.000, 0.7333, 0.1667, 0.0033, 0.0150];
    p_Cdc = 0.3; % Converter capacitor

    p_Ri= 0.0;
    p_Rg= 0.0;
    p_Rc= p_FLTR(3);
    p_Li= p_FLTR(4);
    p_Lg= p_FLTR(5);
    p_Cf = p_FLTR(6);
    %

    % BEGINS: LCL Filter initialization

    % Output of the WTG at the generator bus
    Spmsg = bus_sln(Pmachs,4)+1i*bus_sln(Pmachs,5);
    % Voltage at the PMSG bus
    vpmsg = bus_sln(Pmachs,2).*exp(1i*bus_sln(Pmachs,3)*pi/180);
    % output current of WTG or current through LCL filter grid side
%inductance
    ipmsg = conj(Spmsg./vpmsg);
    % Angle of voltage at the generator bus
    p_Theta = angle(vpmsg);

    % Voltage across R and C in the filter
    vnode = vpmsg+ipmsg.*(p_Rg+1i*p_Lg);
    % Current through the capacitor
    icf = vnode./(p_Rc-1i/(p_Cf));
    % Voltage across capacitor
    vcf = icf.*(-1i/(p_Cf));
    % Current through inverter side inductor
    iLi = ipmsg+icf;
    % Voltage at the inverter side of filter/ voltage output of GSC
    vLi = vnode+iLi*(p_Ri+1i*p_Li);

    % State variables in the program
    p_iiq = real(iLi);   p_iid = imag(iLi);
    p_igq = real(ipmsg); p_igd = imag(ipmsg);
    p_vcq = real(vcf);   p_vcd = imag(vcf);
    p_viq = real(vLi);   p_vid = imag(vLi);


    % BEGINS: Initialisation of grid side converter

    % Parameters for GSC controller model

    % Current and voltage in new reference inside controller
    p_vi_dash = (p_viq+1i*p_vid).*exp(-1i*p_Theta);
    p_ii_dash =  (p_igq+1i*p_igd).*exp(-1i*p_Theta);

p_GSC_IL1_iv = real(p_vi_dash); p_GSC_IL1_kp = 0.3; p_GSC_IL1_ki= 200;
```

Script 6.7 Program for find_pmsg_state_initial_conditions.m.

```
p_GSC_IL2_iv = imag(p_vi_dash); p_GSC_IL2_kp = 0.3; p_GSC_IL2_ki= 200;
p_GSC_OL1_iv = real(p_ii_dash); p_GSC_OL1_kp = -22; p_GSC_OL1_ki = -870;
p_GSC_OL2_iv = imag(p_ii_dash); p_GSC_OL2_kp = 0;  p_GSC_OL2_ki = -60;


    % Reactive power reference GSC controller
    p_Qfilter = imag(Spmsg);

    % B2B capacitor
    p_VDC = 1.5;

    % Initialization of generator states and turbine
    % Assuming no loss in the converter, generator output is equal to
%GSC
    % output
    Pg = real(vLi.*conj(iLi));

    % wind speed
    p_vw = (Pg*p_base*1e6./(0.5*1.225*pi*p_bl^2*0.48)).^(1/3);
    % turbine speed
    p_wt = p_Sw*p_lambda_opt/(p_bl*p_wt_rated);
    p_beta = 0;
    % assuming d-axis current is equal to zero, q-axis current and
%voltage
    % are
    p_isq = Pg./(p_wt.*p_psi);
    p_vsq = Pg./p_isq;

    % d-axis current and voltage are
    p_isd = 0;
    p_vsd = p_Lq.*p_isq.*p_wt - p_ra*p_isd;


    % Initialisation of machine side converter
    p_MSC_IL1_kp = -90;
    p_MSC_IL1_ki = -1000;
    p_MSC_IL1_iv = p_vsq;

    p_MSC_IL2_kp = -90;
    p_MSC_IL2_ki = -1000;
    p_MSC_IL2_iv = p_vsd;
```

Script 6.7 (*continued*).

Table 6.6

Eigenvalues	Frequency (Hz)	Damping ratio (%)
$-58837 \pm 62318i$	9918	68.65
$-16236 \pm 17285i$	2751	68.46
$-517.78 \pm 805.68i$	128	54.06
$-50.23 \pm 386.62i$	61.5	12.88
-64.27	0	100.0
$-23.65 \pm 35.28i$	5.61	55.68
$-378.37 \pm 2.90i$	0.46	99.99
-0.69	0	100.0
$-11.45 \pm 0.09i$	0.014	99.99

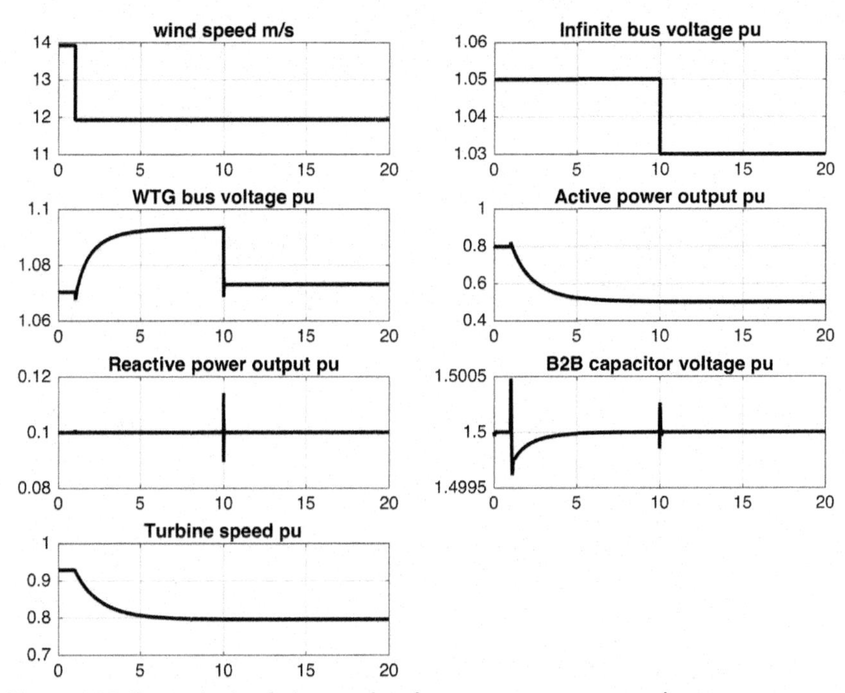

Figure 6.28 Dynamic simulation results of permanent magnet synchronous generator (PMSG)—single machine infinite bus (SMIB) system.

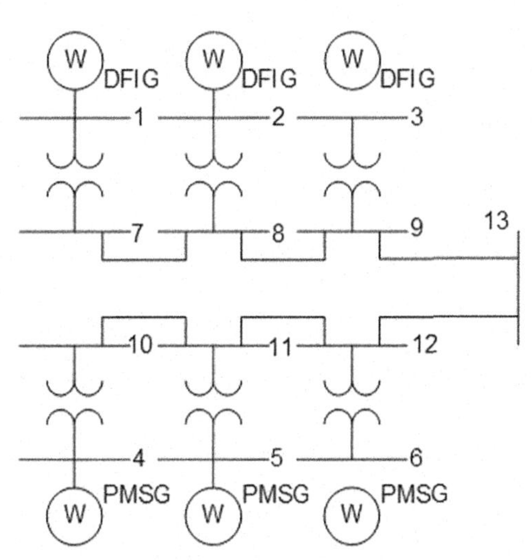

Figure 6.29 Schematic diagram of an example wind farm.

```
clear all

bus = [...
    01 1.00  0.00  0.80  0.26  0.00   0.00  0.00  0.00  3;
    02 1.00  0.00  0.95  0.31  0.00   0.00  0.00  0.00  3;
    03 1.00  0.00  0.90  0.29  0.00   0.00  0.00  0.00  3;
    04 1.00  0.00  0.85  0.28  0.00   0.00  0.00  0.00  3;
    05 1.00  0.00  0.90  0.29  0.00   0.00  0.00  0.00  3;
    06 1.00  0.00  0.95  0.21  0.00   0.00  0.00  0.00  3;
    07 1.00  0.00  0.00  0.00  0.00   0.00  0.00  0.00  3;
    08 1.00  0.00  0.00  0.00  0.00   0.00  0.00  0.00  3;
    09 1.00  0.00  0.00  0.00  0.00   0.00  0.00  0.00  3;
    10 1.00  0.00  0.00  0.00  0.00   0.00  0.00  0.00  3;
    11 1.00  0.00  0.00  0.00  0.00   0.00  0.00  0.00  3;
    12 1.00  0.00  0.00  0.00  0.00   0.00  0.00  0.00  3;
    13 1.00  0.00  0.00  0.00  0.00   0.00  0.00  0.00  1];

line = [
    01  07 0.010  0.10  0.000 0.0 0.0;
    02  08 0.010  0.10  0.000 0.0 0.0;
    03  09 0.010  0.10  0.000 0.0 0.0;
    04  10 0.010  0.10  0.000 0.0 0.0;
    05  11 0.010  0.10  0.000 0.0 0.0;
    06  12 0.010  0.10  0.000 0.0 0.0;
    07  08 0.0048 0.080 0.0010 0.0 0.0;
    08  09 0.0048 0.080 0.0010 0.0 0.0;
    09  13 0.0048 0.080 0.0010 0.0 0.0;
    10  11 0.0048 0.080 0.0010 0.0 0.0;
    11  12 0.0048 0.080 0.0010 0.0 0.0;
    12  13 0.0048 0.080 0.0010 0.0 0.0];

Y = form_Ymatrix(bus,line);
[bus_sln, flow] = power_flow(Y, bus, line);
Znet = inv(Y);
vinf = bus_sln(13,2)*exp(1i*bus_sln(13,3)*pi/180);
ZA = Znet(1:end-1,1:end-1); ZB = Znet(1:end-1,end);
ZC = Znet(end,1:end-1); ZD = Znet(end,end);  % Znet = [Za Zb; Zc Zd]
%%%%%%%%%%%%%%%%%%%%%%%%%%%%%%%%%%%%%%%%%%%%%%%%%%%


Pmachs = [1,2,3]';      % Buses where PMSG is connected
Dmachs = [4,5,6]';       % Buses where DFIG is connected
Omega = 2*pi*50;
if size(Pmachs,1)

    find_pmsg_state_initial_conditions
    pmsg_mult = zeros(size(bus_sln,1)-1,size(Pmachs,1));
    pmsg_mult(Pmachs,:) = eye(size(Pmachs,1));
end


if size(Dmachs,1)

    find_dfig_state_initial_conditions
    dfig_mult = zeros(size(bus_sln,1)-1,size(Dmachs,1));
    dfig_mult(Dmachs,:) = eye(size(Dmachs,1));
end
```

Script 6.8 Program to initialize wind farm model states.

an infinite bus through a network. However, in this system, six WTGs are connected to an infinite bus through a network. Let us discuss how to represent the network in this case.

6.9.1 Network representation

Let V_b, Z_{net} and I_b be vector (13 × 1) of bus voltages, impedance matrix (13 × 13) and vector (13 × 1) of current injection at the buses, respectively. Then,

$$V_b = Z_{net} I_b \tag{6.44}$$

$$\begin{bmatrix} V_1 \\ \cdot \\ \cdot \\ V_{12} \\ V_{13} \end{bmatrix} = \begin{bmatrix} Z_{1,1} & \cdot\ \cdot & Z_{1,12} & Z_{1,13} \\ \cdot & \cdot\ \cdot & \cdot & \cdot \\ \cdot & \cdot\ \cdot & \cdot & \cdot \\ Z_{12,1} & \cdot\ \cdot & Z_{12,12} & Z_{12,13} \\ Z_{13,1} & \cdot\ \cdot & Z_{13,12} & Z_{13,13} \end{bmatrix} \begin{bmatrix} I_1 \\ \cdot \\ \cdot \\ I_{12} \\ I_{13} \end{bmatrix} \tag{6.45}$$

The current injection at buses 1–6 can be obtained from WTG models. There is no current injection at buses 7–12. As Bus 13 is an infinite bus, we do not know the current injection (I_{13}) at Bus 13, but we know the voltage (V_{13}). Let us rewrite the equation as

$$\begin{bmatrix} V_b \\ V_{13} \end{bmatrix} = \begin{bmatrix} Z_A & Z_B \\ Z_C & Z_D \end{bmatrix} \begin{bmatrix} I_b \\ I_{13} \end{bmatrix} \tag{6.46}$$

From (6.46),

$$I_{13} = Z_D^{-1}(V_{13} - Z_C I_b) \tag{6.47}$$

Using (6.47), we can write

$$V_b = \left(Z_A - Z_B Z_D^{-1} Z_C\right) I_b + Z_B Z_D^{-1} V_{13} \tag{6.48}$$

In the program, the matrix Znet is divided into four sub matrices: ZA, ZB, ZC and ZD. Simulink representation of the network using Eq. (6.48) is shown in Fig. 6.30.

6.9.2 Wind farm simulink model

The Simulink model for the wind farm simulation is shown in Fig. 6.31. Now develop the wind_farm_init as given in Script 6.8 and wind_farm_model

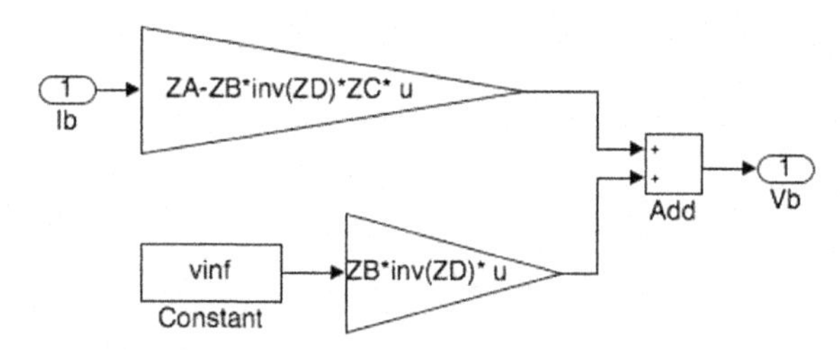

Figure 6.30 Simulink representation of the wind farm network.

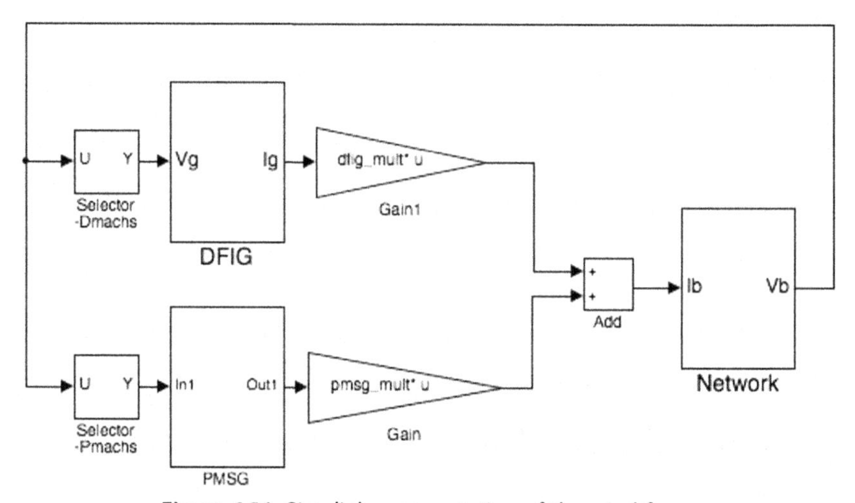

Figure 6.31 Simulink representation of the wind farm.

as shown in Fig. 6.31. The output of the network block is vector of size 12 representing bus voltages at buses 1–12. The selector elements from the 'Signal Routing' library of the Simulink are used to select appropriate voltages for the DFIG and PMSG blocks. The variables Dmachs and Pmachs are used to specify the bus numbers. Make changes in the block parameter dialogue of the selector elements: index option: Index vector (dialogue), Index: Dmachs or Pmachs and Input port size: 12.

Now the output of DFIG and PMSG blocks are current vector of size equal to the number of DFIG and PMSG in the network. We need to convert them into a vector of 12 elements representing current injections at buses 1–12. The variables dfig_mult and pmsg_mult are used to modify the current vector. In the 'block parameter' dialogue of the dfig_mult and pmsg_mult, select Matrix(K*u) for the multiplication option. A

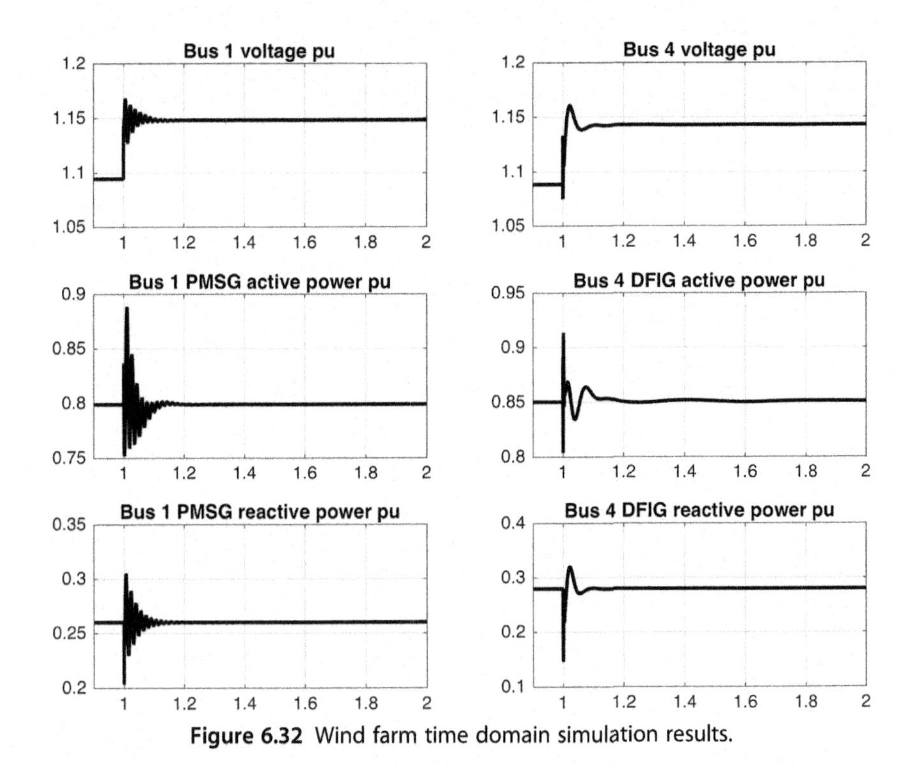

Figure 6.32 Wind farm time domain simulation results.

response to step change in the infinite bus voltage from 1.0 pu to 1.05 pu at time = 1sec is shown in Fig. 6.32 for comparison.

References

Abad, G., Lopez, J., Rodriguez, M., Marroyo, L., Iwanski, G., December 2011. Doubly Fed Induction Machine: Modeling and Control for Wind Energy Generation. Wiley-IEEE Press.

Araujo, S.V., Engler, A., Sahan, B., Antunes, F.L.M., 2007. LCL filter design for grid-connected NPC inverters in offshore wind turbines. In: Proc. 7th Int. Conf. Power Electron., Daegu, pp. 1133–1138.

Kim, H.-W., Kim, S.-S., Ko, H.-S., 2010. Modeling and control of PMSG-based variable-speed wind turbine. Electr. Power Syst. Res. 80 (1), 46–52 (PMSG model reference).

Li, S., Haskew, T.A., Xu, L., 2010. Conventional and novel control designs for direct driven PMSG wind turbines. Electr. Power Syst. Res. 80 (3), 328–338.

Mei, F., 2008. Small-Signal Modelling and Analysis of Doubly-Fed Induction Generators in Wind Power Applications. PhD thesis. Imperial College, London, United Kingdom.

Rosyadi, M., Muyeen, S.M., Takahashi, R., Tamura, J., 2012. New controller design for PMSG based wind generator with LCL-filter considered. In: Proc. XXth Int. Conf. Elect. Mach. Marseille, pp. 2112–2118.

Rosyadi, M., Umemura, A., Takahashi, R., Tamura, J., Uchiyama, N., Ide, K., 2014. A new simple model of wind turbine driven Doubly-Fed Induction Generator for dynamic analysis of grid connected large scale wind farm. In: 3rd Renewable Power Generation Conference (RPG 2014), Naples, pp. 1–6 [Open Access].

Wu, B., Lang, Y., Zargari, N., Kouro, S., August 2011. Power Conversion and Control of Wind Energy Systems. Wiley-IEEE Press.

Wu, F., Zhang, X.P., Godfrey, K., Ju, P., September 2007. Small signal stability analysis and optimal control of a wind turbine with doubly fed induction generator. IET Gen. Trans. Distrib. 1 (5), 751–760.

CHAPTER SEVEN

Modelling of solar generation

This chapter covers current methods of PV modelling and associated challenges. A model for distributed PV generation (WECC, Generic solar photovoltaic system dynamic simulation model specification, Sep. 2012) is presented in detail, with the associated Matlab and Simulink modelling.

Wind generation and photovoltaic (PV) generation are in many countries the top two renewable generation technologies. Still, the modelling of PV is not as well discussed and developed compared with its wind generation counterpart. Utilities, however, need to account for the PV generation in their systems for power flow as well as dynamic studies (Ellis, n.d.). Some of the factors slowing the development of comprehensive PV models are alike with those seen in the wind sector, namely the confidentiality around vendor specific models, which leads to a lack of cross-validation of modelling standards and a lack of comprehensive generic models. The Western Electricity Coordinating Council (WECC) has therefore developed two generic PV generator models, one for very large PV plants connected at transmission level and a second model for distributed PV generation, connected at distribution level, suitable for transmission level studies. Both models are well documented and supplemented by a number of reports (WECC, WECC solar plant dynamic modeling guidelines, Apr. 2014) (WECC, Central Station Photovoltaic Power Plant Model Validation guideline, Mar. 2015). The use of these models has become widespread in literature. This chapter presents in detail how to include the distributed PV model in your Matlab/Simulink simulations.

7.1 Description of solar generation

Installations of PV s vary widely, from individual installations of PV panels on an increasing number of homes to large PV power stations. For example countries such as China, India and USA have added capacity into the 10s of GW during 2018 alone (International Energy Agency, 2019).

An important point to note, as a power system engineer interested in the impact solar PV has on the network, is unlike all other generation types, PV generation has no moving mass and hence no natural form of inertia.

Simulation of Power System with Renewables
ISBN: 978-0-12-811187-1
https://doi.org/10.1016/B978-0-12-811187-1.00007-X
© 2020 Elsevier Inc.
All rights reserved.

All purely DC connected generation is lacking any natural expression of inertial response, such as Type IV wind generation. Some of these technologies have been equipped with a fast control response to exhibit a power output change in response to the rate of change of frequency or a frequency change to emulate the inertial response behaviour. Similarly, any response to frequency changes from PV generators is provided through control systems.

Another phenomena that is particular for photovoltaic generation is the power generation bell curve during the day. As the sun rises, PV panels can produce an increasing amount of electricity. As the sun sets, the power prodution decreases again. This leads to the characterisitc bell shape. As clouds pass over a solar panel, the PV panel is shielded from the radiation and power output plumets and rises at extremly fast rates. Why not use a small solar panel to observe this effect for yourself or search the literature for this plot?

7.2 Modelling solar power generators

In general, the PV system can be separated into different components (or subsystems) for modelling purposes, some of which are well known to the modern power system engineer. As shown in Fig. 7.1, the model contains an inverter subsystem, which converts the generated DC power into AC, according to a chosen control scheme and converter topology.

The main subsystem that is particular to this technology is the PV array model, responsible to represent the energy conversion from solar irradiance to DC output power. The generated DC power is a function of solar irradiance. The power flows to the inverter, which operates at a certain DC voltage, with the available DC current for any particular value of ambient temperature (Ellis, n.d.).

The total power P_{DC} produced by the PV array depends not only on the solar irradiance but also on the voltage and temperature (Ellis, n.d.)

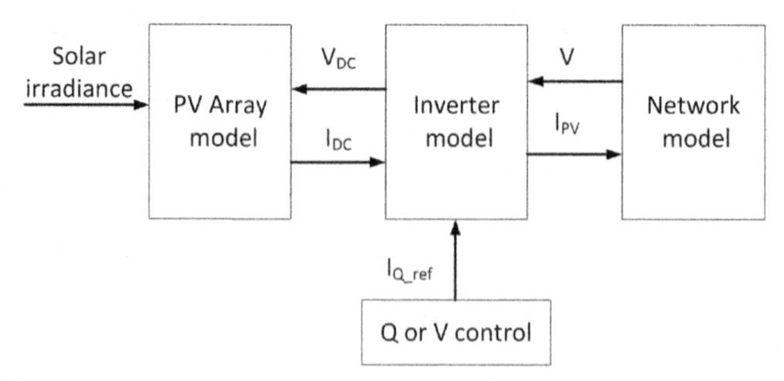

Figure 7.1 Diagrammatic view of physical photovoltaic modelling.

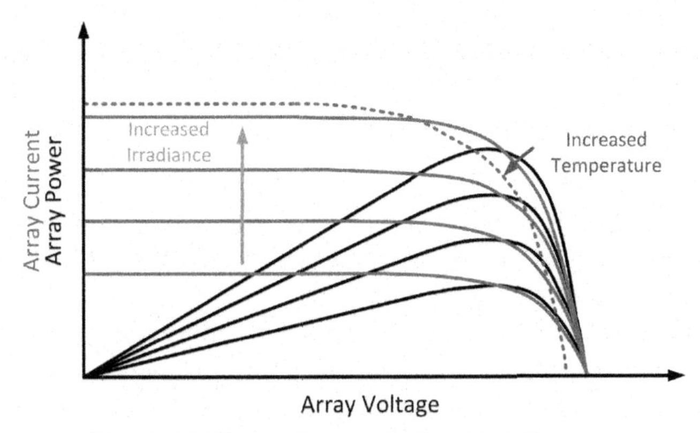

Figure 7.2 Photovoltaic array characteristic curve.

(Lawson, n.d.). Datasheets for PV panels typically include such curves for the relationship of current and power with voltage. A qualitative view of such a typical relationship is shown in Fig. 7.2. While the available power P_{DC} is an intermittent function of the solar irradiance available, the reactive power generated can be controlled by the inverter, in either a reactive power, power factor or voltage control mode. On the AC side, the grid at the point where the PVs are connected has a certain voltage. The inverter injects the current I_{PV}, delivering the power generated by the PV array to the grid.

The operation of a PV cell can further be represented by an equivalent circuit, where this is required. The photocurrent of the cell can be linked to the terminal voltage and terminal current of the cell, via an antiparallel diode and parallel as well as series resistance (Yashodhan, 2014), (Ramachandran, 2005) (Tan, 2014).

As is the case with wind farm modelling, there are modelling options. In one method, each individual wind turbine is modelled in detail, and the wind farm is represented by an array of wind turbines connected via the wind farm grid. Another modelling assumption that is frequently made is clustering of wind turbines with similar operating behaviour into equivalent machines to reduce the number of wind turbines that are explicitly modelled. A third approach is to model the whole wind farm by one dynamic equivalent model. This model avoids the detailed representation of wind turbines all together and is a method of choice for certain applications, where the dynamics of the wind farm power injection is of concern, while the detailed turbine dynamics are not. For large-scale PV integration, this type of model is the most commonly used, and this is the model we will study in more detail in this book.

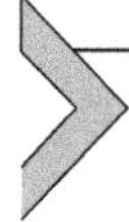

7.3 Western Electricity Coordinating Council generic model

As mentioned, the WECC has developed two generic PV generator models, one for very large PV plants connected at transmission level and a second model for distributed PV generation connected at distribution level, suitable for transmission level studies. In this chapter, we implement the distributed PV model in Matlab/Simulink.

The aim of the models is to represent the dynamics at the point of interconnection, rather than inside the PV plant itself. The models do not include a PV array model, which links solar irradiance with voltage and temperature, as described above. Rather, available solar power is assumed to be a certain value, and the focus of the model is the study of electrical disturbances. The recommended integration time step is 1–10 ms with a typical duration of simulation in the 30 s range. Models are able to represent phenomena of up to 10Hz. The model should only be used down to 25% of rated power and an SCR of at least 2 at the point of interconnection (WECC, Generic solar photovoltaic system dynamic simulation model specification, Sep. 2012). The WECC PV models may produce a very fast voltage spike of less than a cycle at the onset of the fault and clearing. These spikes should be neglected, as they do not represent the physical behaviour. Furthermore, these generic models are not intended for fault ride through studies (WECC, WECC solar plant dynamic modeling guidelines, Apr. 2014).

7.4 Case study: photovoltaic system model

In this case study, we use the same four-machine system as before, with distributed PV generation feeding into Bus 7, via a substation unit transformer, equivalent feeder impedance and generation step-up transformer (WECC, WECC solar plant dynamic modeling guidelines, Apr. 2014) as shown in Fig. 7.3.

For this simulation, we add three additional buses, Bus 12, 13 and 14, to our four-machine bus and line data from Chapter 3 as shown in Script 7.1.

Copy Script 7.2 and save it as PV_values.m and run it as well. This script contains all the parameters required for the PV system model.

With the full initialization of this four-machine plus PV, 14-bus system in place, our attention turns toward the Simulink model for the PV system. Fig. 7.4 shows how the PV model is integrated with the four-machine model containing a balanced three-phase fault from Chapter 3 named

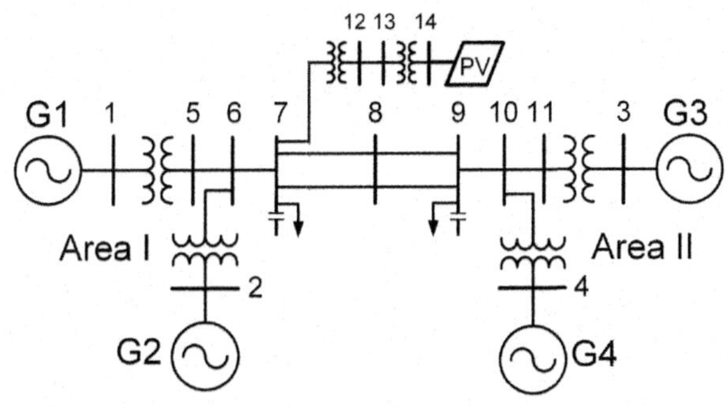

Figure 7.3 Case study – four-machine system with distributed photovoltaic generation.

```
bus = [...

as before

12 1.0 0.0 0.0 0.0    0.0 0.0 0.0 0.0 3;
13 1.0 0.0 0.0 0.0    0.0 0.0 0.0 0.0 3;
14 1.0 0.0 0.2 0.0125 0.0 0.0 0.0 0.0 3];


line = [...

as before

07 12 0      0.05   0    1.0 0.0;
12 13 0.0075 0.0125 0.02 1.0 0.0;
13 14 0      0.025  0    1.0 0.0];
%parameters p.16 WECC Solar plant Dynamic modelling guidelines
```

Script 7.1 Photovoltaic addition to four-machine data (WECC Sept 2012).

two_area_synch_model.slx. The PV model receives voltage as input from the network model and the required frequency input from a nearby synchronous machine. To use the frequency signal from a generator, simply make del_w and output of the synchronous machine model and feed it into the PV model. The output current from the PV generation is added to the current injections of all other generators, to calculate the current injections for the network.

Fig. 7.5 shows the distributed PV model. The model allows for a droop setting, to alter the real power for the PV system to help with frequency control. Equally, the model allows for the reactive power to be altered for the purposes of voltage control. Inside the PV model, the reference frame is aligned with the terminal voltage, such that reference currents can be calculated from voltage magnitude and real and reactive power, respectively. The

```
%From WECC documentation
Pqflag=1; % Priority to reactive current (0) or active current
%(1)
Xc=0; %Line drop compensation reactance (pu on mbase)
Qmx=0.328; %Maximum reactive power command (pu on mbase)
Qmn=-0.328; %Minimum reactive power command (pu on mbase)
V0=0.9; %Lower limit of deadband for voltage droop response
%(pu)
V1=1.1; %Upper limit of deadband for voltage droop response
%(pu)
Dqdv=0; %Voltage droop response characteristic
fdbd=-99; %Overfrequency deadband for governor response (pu
%deviation)
Ddn=0; %Down regulation droop gain on mbase
Imax=1.1; % Apparent current limit (pu on mbase) 1.0 to 1.3
Vt0=0.88; %Voltage tripping response curve point 0 (pu)
Vt1=0.9; %Voltage tripping response curve point 1 (pu)
Vt2=1.1; %Voltage tripping response curve point 2(pu)
Vt3=1.2; %Voltage tripping response curve point 3 (pu)
vrrecov=0; %Voltage tripping is latching (0) or partially
%self-resetting (>0 and <=1)
Ft0=59.5; %Frequency tripping response curve point 0 (hz)
Ft1=59.7; %Frequency tripping response curve point 1 (hz)
Ft2=60.3; %Frequency tripping response curve point 2 (hz)
Ft3=60.5; %Frequency tripping response curve point 3 (hz)
frrecov=0; %Frequency tripping is latching (0) or partially
%self-resetting (>0 and <=1)
Tg=0.02; %Inverter current lag time constant(s)
P_ext=0; %Supplemental active power signal (pu on mbase; zero
%unless written by external model)
Freq_ref=60; %Initial terminal voltage frequency %
%Documentation says deviation, but seems to be actual.

Vt_set=abs(vol(14));

%Additional parameters
Mbase=25; %MVA
Sbase=100; %MVA
SelectPV=zeros(1,14);
SelectPV(1,14)=1;
ExpandPV=zeros(14,1);
ExpandPV(14,1)=1;
Baseconv=Mbase./Sbase;

Q_ref=0.05; % ##Choose suitable value Initial reactive power
%(pu on mbase, from power flow solution)
P_ref=0.8; % ##Choose suitable value Initial active power (pu
%on mbase, from power flow solution)

I_init=conj(P_ref+1i*Q_ref)/vol(14);
Iq_init=imag(I_init);
Id_init=real(I_init);
```

Script 7.2 PV_values.m (WECC Sept 2012).

PV reference current to current stage is subject to operation limits and delays. The current of the PV model needs to be converted to system base and aligned according to the system reference. The PV model also includes several flags for tripping of the modules in case of over or under frequency or voltage.

Let us discuss these seven subsystems in more detail, moving from top to bottom and left to right. Fig. 7.6 shows that we have selected the frequency of the second generator and converted it from the per unit difference to the nominal speed back to its synchronous speed in Hz. We have chosen the second machine, for its location close to the PV system.

Fig. 7.7 shows the subsystem for the PV systems terminal voltage. Firstly, only voltages at PV terminal buses are selected. The PV system model is aligned with its terminal voltage, such that only voltage magnitude is required inside the model. The voltage angle is required for the back conversion to the network reference frame. The voltage also has a delay block to break algebraic loops, as discussed in Chapter 5.

Fig. 7.8 shows that the total real power the PV system is generating is made up of the reference power, an additional droop setting in case of frequency control and the option for a supplemental active power signal P_ext (zero by default). The droop setting includes a dead zone block, with the starting value set as fdbd and the end value set to a very high value, here 10^20.

The reactive power is set to the reference value with the option to adjust the reference value to include a voltage droop or line drop compensation with the reactance value X_C. The dead zone block has a starting value of V0*Dqdv and an end value of V1*Dqdv. The saturation has an upper limit of Qmx-Q_ref and a lower limit of Qmn-Q_ref (Fig. 7.9).

The PV generation will trip during low voltage, high voltage and low or high frequency conditions. This tripping happens gradually, with a reduction of connected PV output, as the situation worsens. Low voltage tripping is initiated from a voltage below Vt1 with total disconnection at a voltage level of Vt0, whereas high voltage tripping is initiated from a voltage level of Vt2 with total disconnection from a voltage level of Vt3, as shown in Fig. 7.10.

Each low or high frequency or voltage condition has its own flag and tripping logic. These flags are combined to an overall instruction to the PV system, as can be seen in Fig. 7.11.

The logic inside the low voltage tripping subsystem is shown in Figs. 7.12–7.16. Fig. 7.12 shows how Vmin is established as the lowest point that Vt reaches over time, with the saturation block limiting Vmin to a lower limit of Vt0 with an arbitrary upper limit set to a large value, here 10^10.

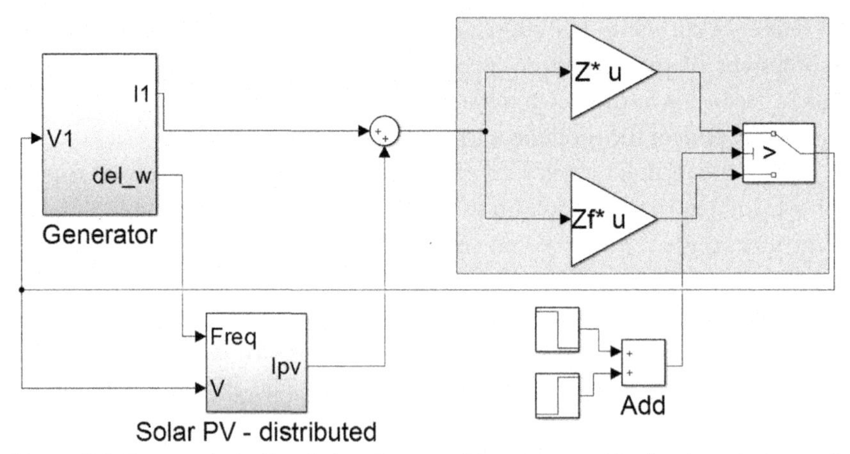

Figure 7.4 Case study in Simulink – four-machine system with distributed photovoltaic generation.

Fig. 7.13 shows the low voltage tripping logic, expressed by a cascade of comparators. If you refer to the WECC documentation (WECC, Generic solar photovoltaic system dynamic simulation model specification, Sep. 2012), you will find this logic expressed as code. Because we wish to implement this in Simulink, this is inconvenient to us, as this function would need to be excetuted at every time step, not only during initilization. Interfacing functions with Simulink code is not trivial and also poses challenges for linearization. A better design choice for us is to directly include this logic through comparators in our model. Let us scan Fig. 7.13 starting from the RHS and working our way back. Refer to the WECC documentation to see if you can understand how the code has been converted to comparators here.

If the current voltage is lower than the low voltage limit Vt0, the low voltage flag is set to zero and our PV generation is disconnected. If both Vt and Vmin are larger than Vt1, the flag is one and no disconnection is initiated by the low voltage loop. If Vt is larger than Vt1 but Vmin is smaller than Vt1, we use the low voltage tripping flag value FvlC shown in Fig. 7.16, when Vt has been below Vt1 but has recovered. The remaining cases are for values of Vt between Vt0 and Vt1. If Vt is recovering, Vt will be larger than Vmin. The flag value used is FvlB, as shown in Fig. 7.15. A decreasing value of Vt between Vt1 and Vt0 has a flag value of FvlA as shown in Fig. 7.14.

The calculation of the low voltage tripping flag value while the voltage is decreasing between Vt1 and Vt0 is shown in Fig. 7.14.

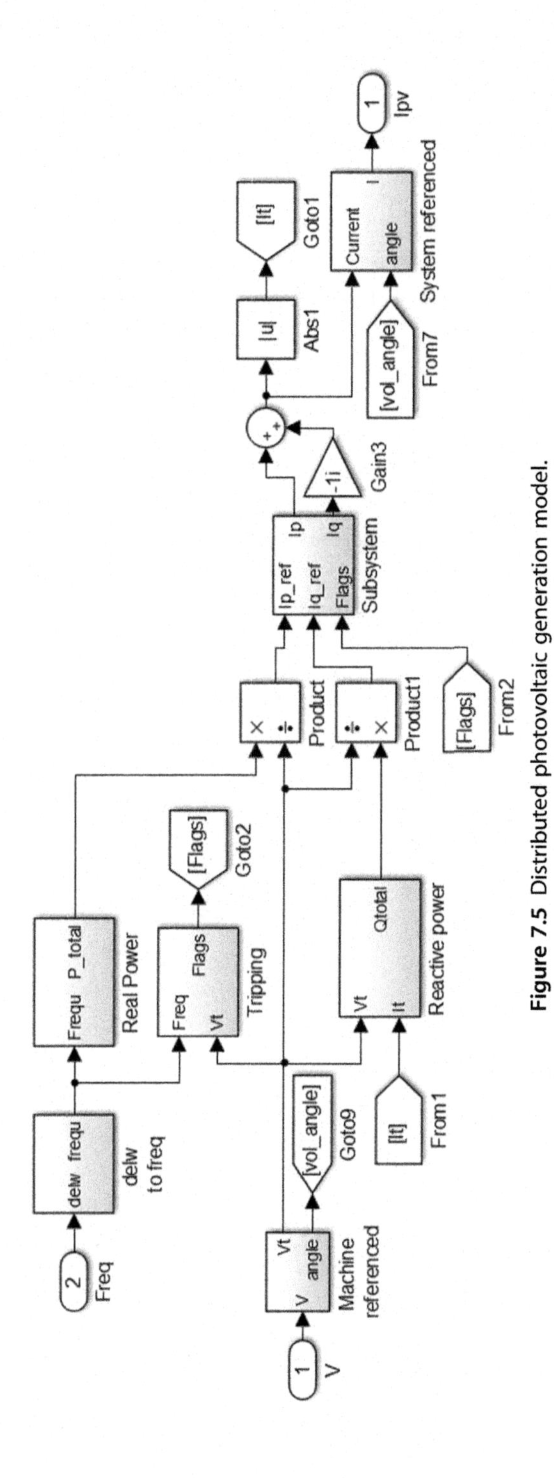

Figure 7.5 Distributed photovoltaic generation model.

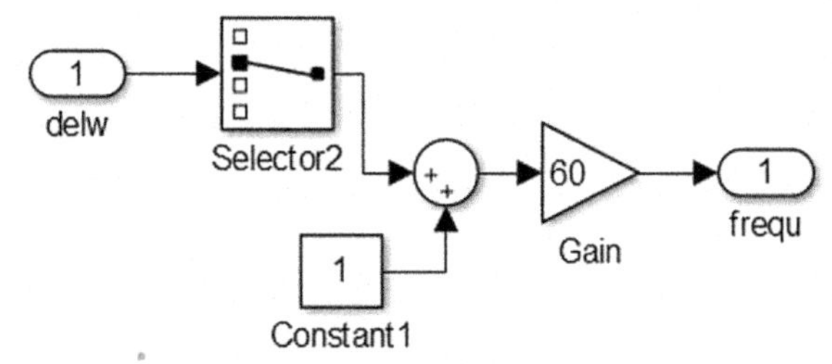

Figure 7.6 Frequency of second synchronous machine in Hz.

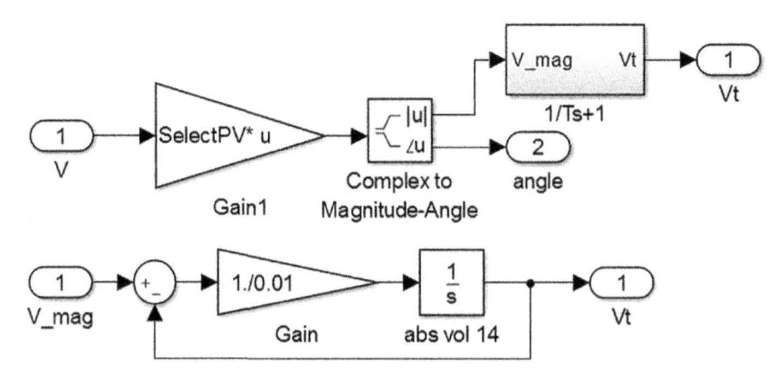

Figure 7.7 Terminal voltage for photovoltaic model.

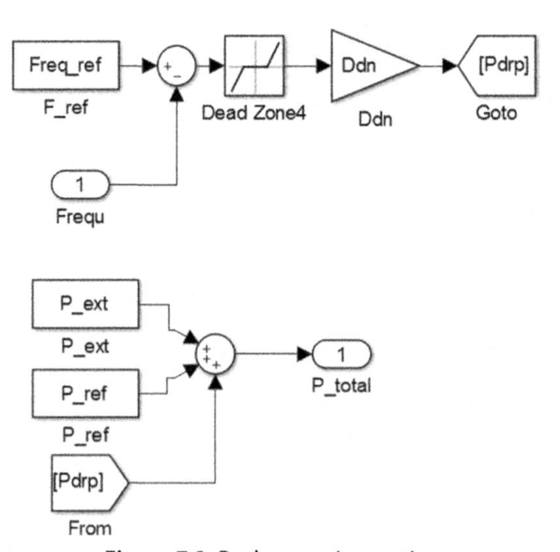

Figure 7.8 Real power instruction.

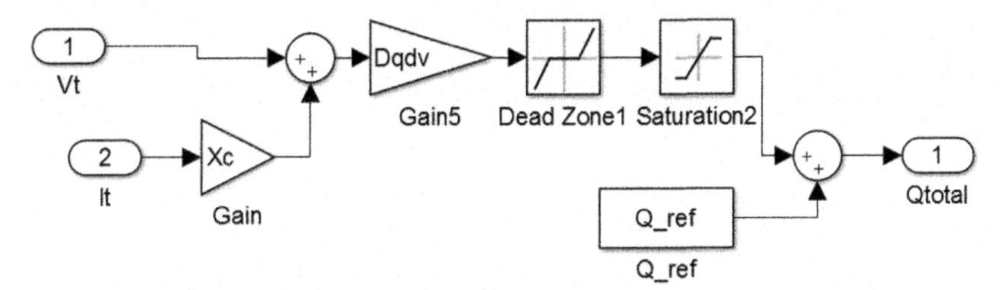

Figure 7.9 Reactive power instruction.

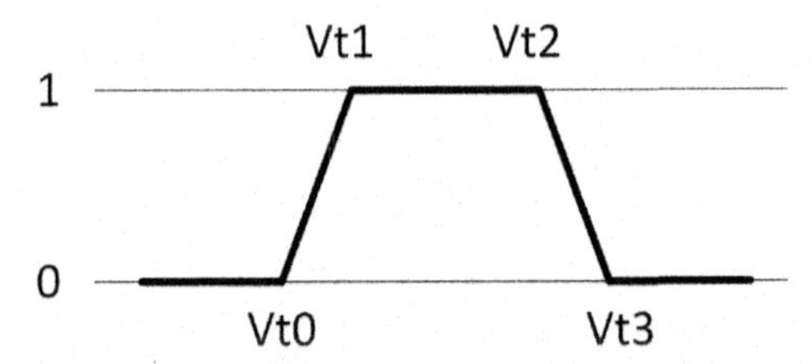

Figure 7.10 Low and high voltage tripping limits Vt0 to Vt3.

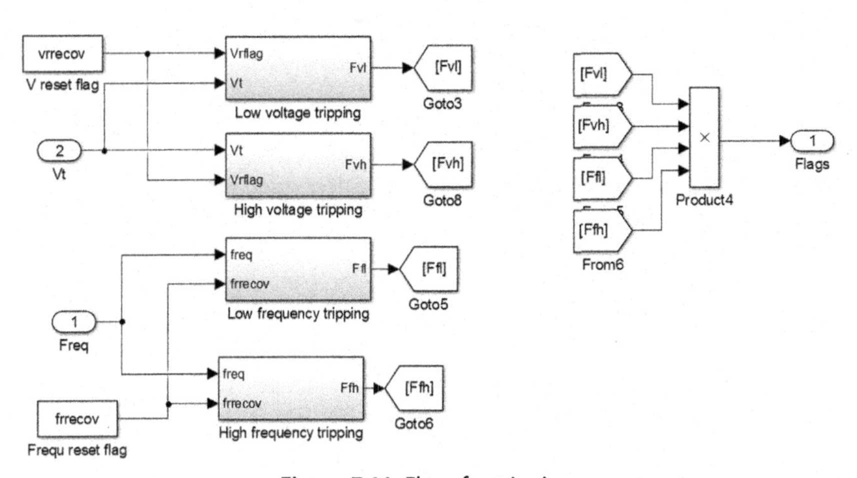

Figure 7.11 Flags for tripping.

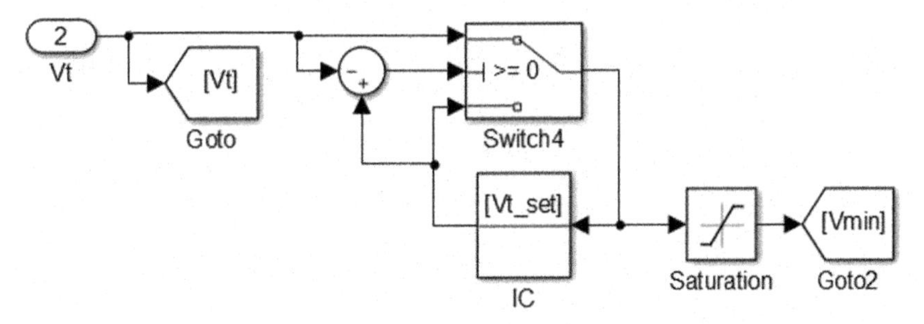

Figure 7.12 Low voltage tripping, tracking lowest voltage Vmin

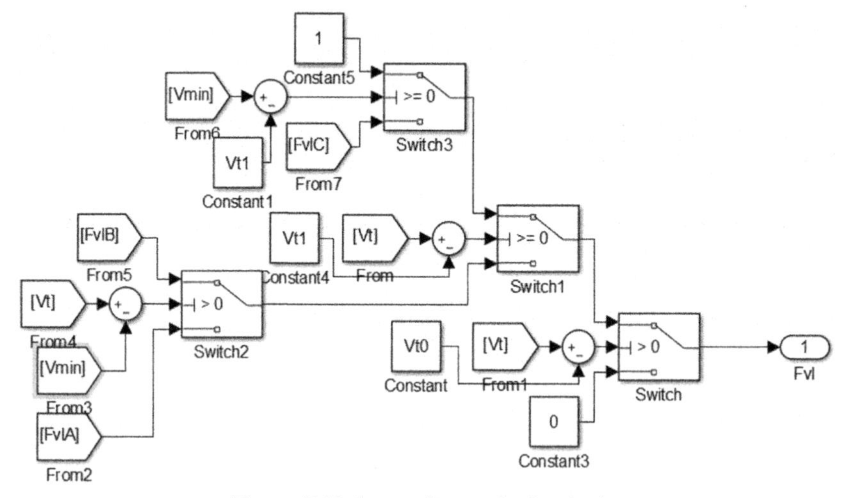

Figure 7.13 Low voltage tripping logic.

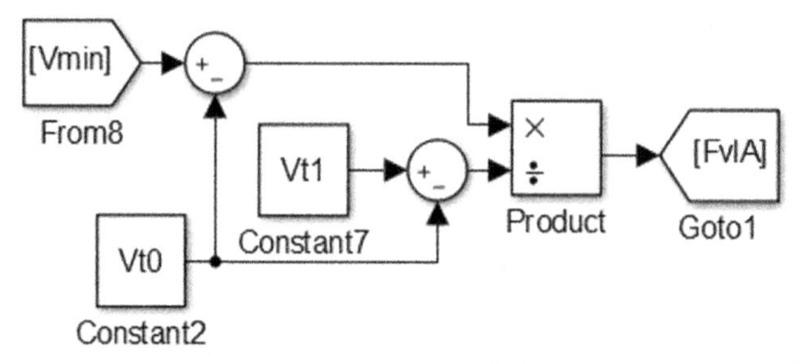

Figure 7.14 Low voltage tripping flag value while decreasing between Vt1 and Vt0.

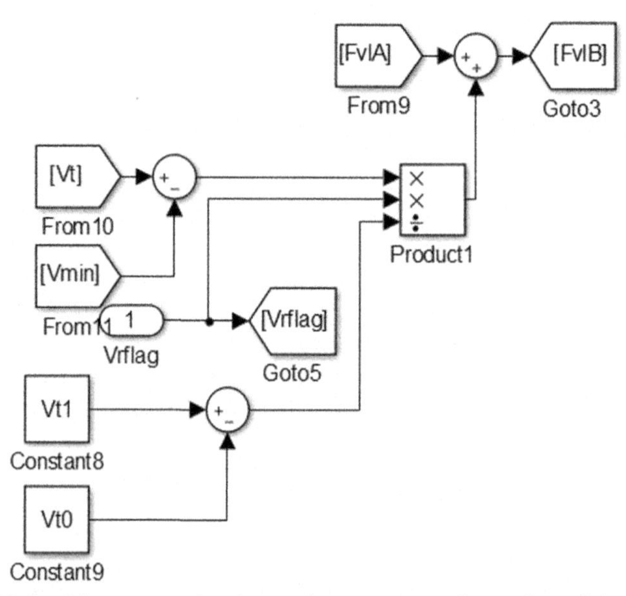

Figure 7.15 Partial reconnection low voltage tripping flag value while recovering above Vmin

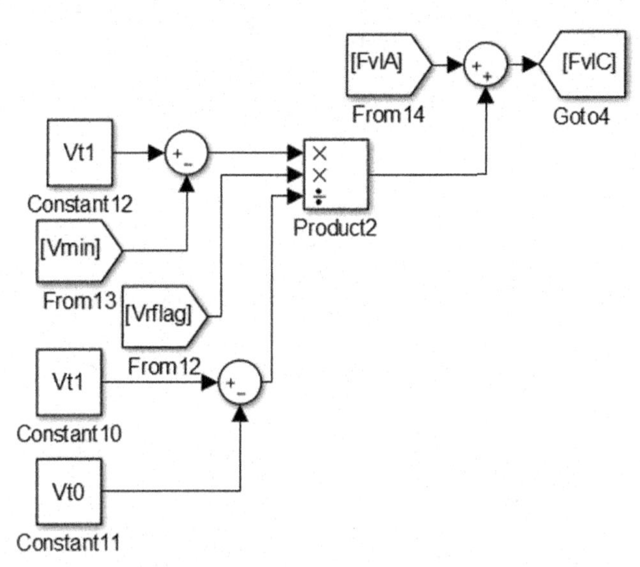

Figure 7.16 Low voltage tripping flag value, when Vt has been below Vt1 but has recovered.

If Vt has not gone below Vt0 and is currently recovering back toward Vt1, Vt will be larger than Vmin. The flag value used is FvlB. The process of obtaining this value is shown in Fig. 7.15.

If Vt is larger than Vt1 but Vmin is smaller than Vt1, we use the low voltage tripping flag value FvlC with the calculation shown in Fig. 7.16. This value is for the case where Vt has been below Vt1 but has recovered.

Fig. 7.17 shows how Vmax is established in the simulation, with the upper limit set to Vt3 and the lower limit set to Vt_set.

Fig. 7.18 is the high voltage tripping logic. Having understood the operation of the low voltage tripping logic, see if you can work out the functionality of the high voltage tripping, which is not dissimilar.

Fig. 7.19 shows the calculation of the high voltage tripping flag value when the voltage is increasing from Vt2 toward Vt3.

Fig. 7.20 shows the calculation of the high voltage tripping flag value while recovering below Vmax.

Fig. 7.21 shows the high voltage tripping flag value, when Vt has been above Vt2 but has recovered.

We will understand the functionality of the tripping logic we implemented through two examples. In the first one, Fig. 7.22, the voltage tripping is latching, meaning it does not automatically recover once tripped. In the second example, Fig. 7.23, the voltage tripping is completely

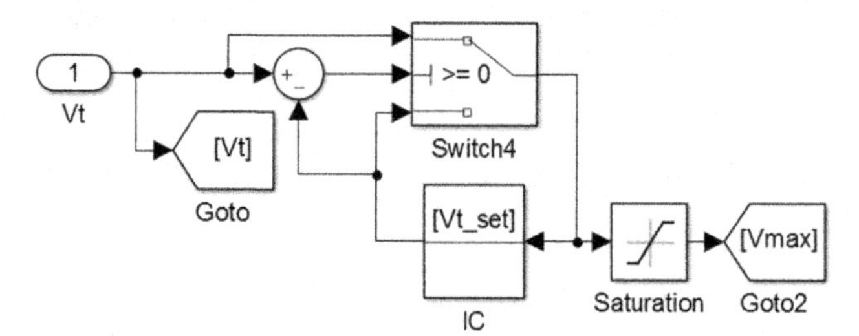

Figure 7.17 High voltage tripping, tracking highest voltage Vmax.

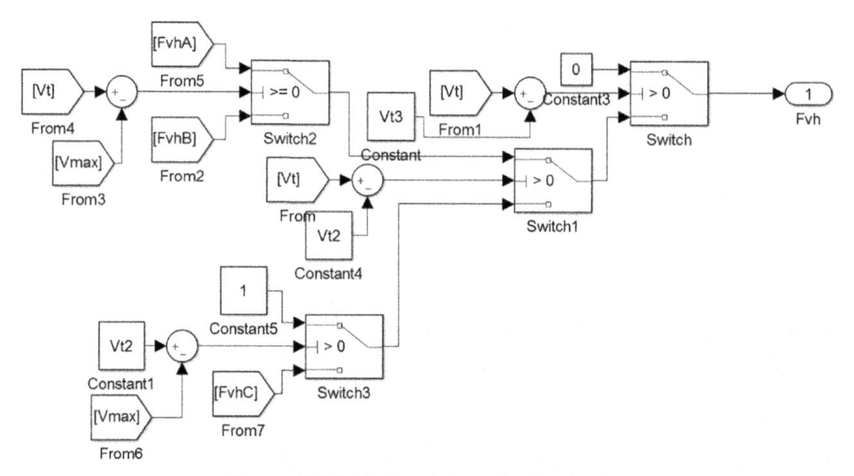

Figure 7.18 High voltage tripping logic.

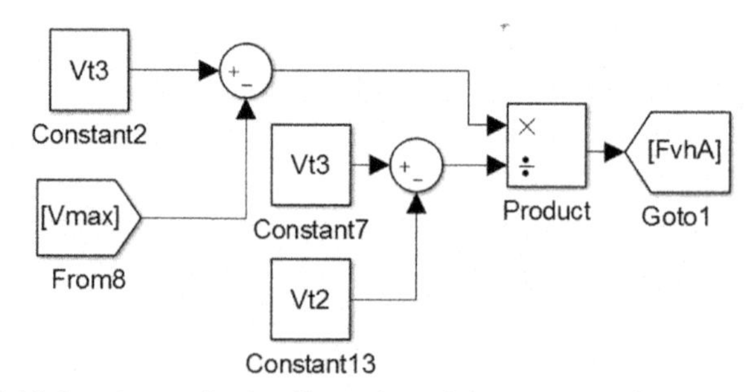

Figure 7.19 High voltage tripping flag value while increasing between Vt2 and Vt3.

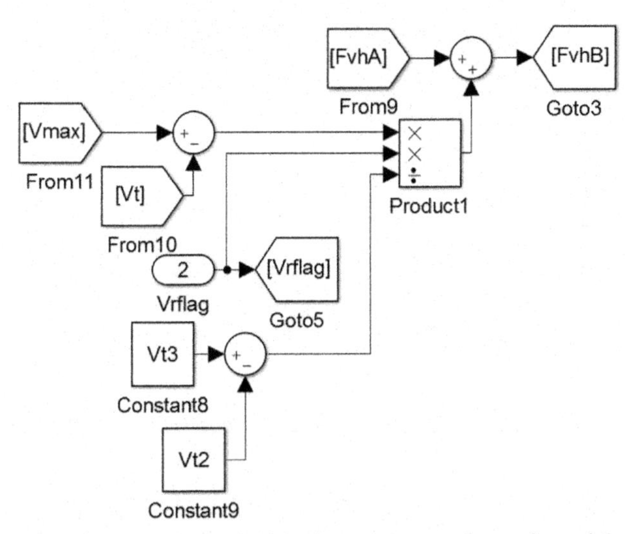

Figure 7.20 Partial reconnection high voltage tripping flag value while recovering below Vmax.

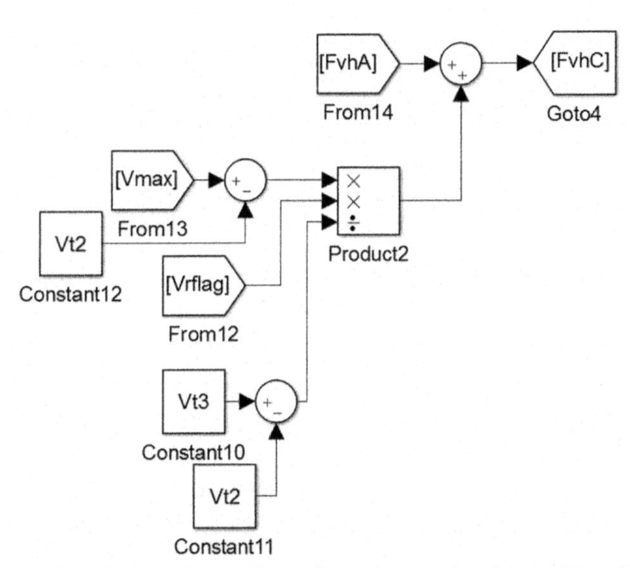

Figure 7.21 High voltage tripping flag value, when Vt has been above Vt2 but has recovered.

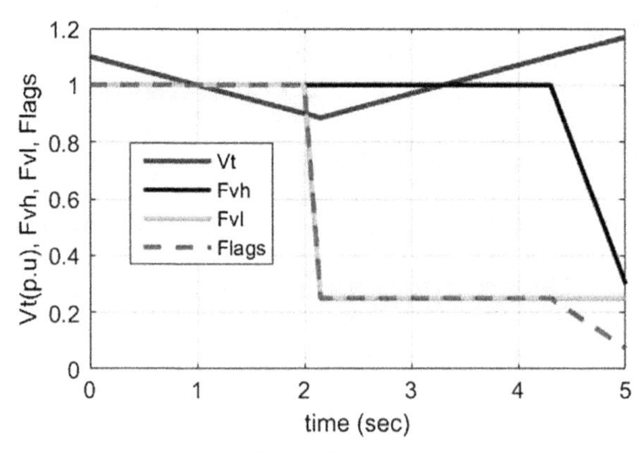

Figure 7.22 Tripping flags with latching voltage tripping, vrrecov = 0.

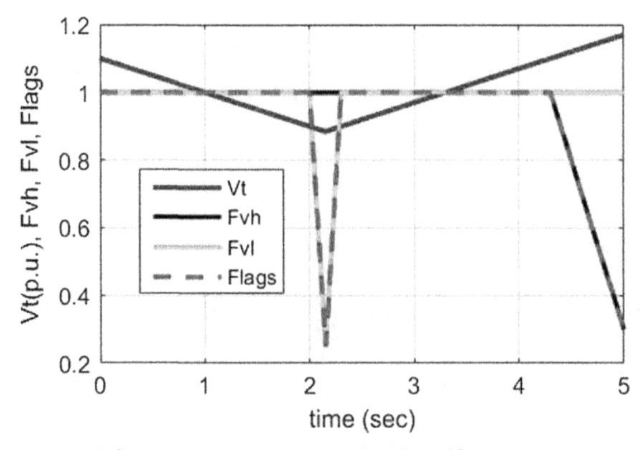

Figure 7.23 Voltage tripping is completely self-resetting, vrrecov = 1.

self-resetting. You may try yourself some other values for vrrecov between 0 and 1 to see what partial self-resetting scenarios look like. We test this section separately from the rest of the PV model, simply by giving a generated voltage input (here we used addition of a decreasing and an increasing voltage ramp) and observing all relevant flag values. At time $t = 0$, we use a ramp rate of -0.1, and at $t = 2.15$, we add a positive ramp rate to this with 0.2, such that from $t = 2.15$ we get a steady increase of $+0.1$.

In Fig. 7.22, the terminal voltage decreases. As it goes down to Vt1, the low voltage flag value Fvl starts to decrease, for partial disconnection of the PV generation. In this scenario, the voltage Vt eventually recovers, but as this is the latching scenario, disconnected PV generation is not reconnected, and the low voltage flag remains at the same value, from the moment Vt is improving. Vt then further increases until it exceeds Vt2. At this point, the high voltage flag value starts to decrease, causing partial disconnection of PV generation. The proportion of PV generation remaining depends on the flags value, which is shown by a dashed line, and it is the simple multiplication of all flag values.

In Fig. 7.23, we go through the same scenario but with a fully self-resetting setting. It can be seen that once Vt recovers back to Vt1, the low voltage flag and hence also the flag value have totally recovered, meaning all PV generation will be automatically reconnected. Once Vt exceeds Vt2, the high voltage flag value decreases along with the flags value. If Vt increases beyond Vt3, total disconnection will occur.

Having understood the concept and implementation of the high and low voltage tripping, you are now able to create the subsystems for low and high frequency tripping using the same logic, renaming variables to use 'F' instead of 'V'.

Fig. 7.24 shows how the current is limited by saturation blocks to stay within permissible limits. The currents injected to the grid may further be reduced through our full or partial disconnection flags for low/high voltage/frequency events. Finally the model contains a delay block, which helps break any algebraic loops and also represents any delays from the power electronics.

Fig. 7.25 shows the details of the saturation block for Ip_lim. Imax is the apparent current limit, and the Pqflag gives priority to reactive current (0) or active current (1).

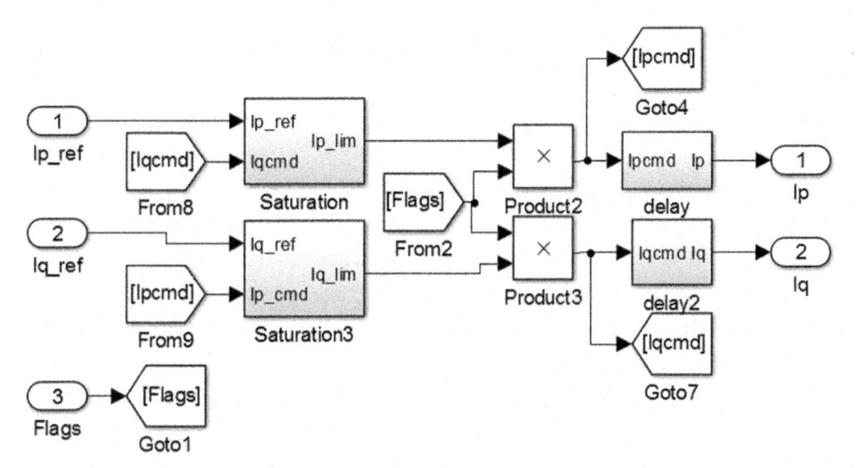

Figure 7.24 Photovoltaic currents from reference values including saturation and delay.

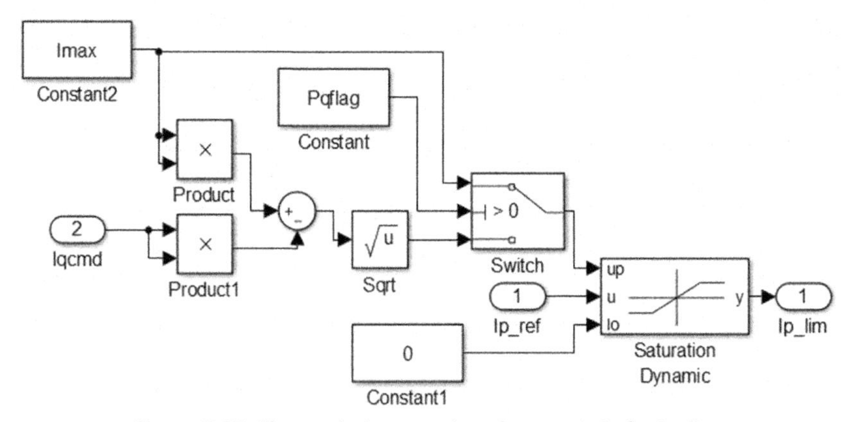

Figure 7.25 Photovoltaic saturation characteristic for Ip_lim.

Fig. 7.26 shows the details of the saturation block for Ip_lim. Imax is the apparent current limit, and the Pqflag gives priority to reactive current (0) or active current (1).

Fig. 7.27 shows the inverter current lag for the p-component with time constant Tg.

The delay block for the q-component is practically identical, so you can implement this as well.

Finally, for the current injected into the grid, we need this in the same reference frame and base units as all other grid quantities. Fig. 7.28 shows how the angle is adjusted back to the network reference frame; the base is converted from machine base to system base, and finally, a current vector is created, which has the same number of elements as the system buses. This vector has zero entries at buses without solar PV and nonzero entries determined by our model at those buses with PV generation.

We will test the behaviour of the distributed PV model using a balanced three-phase fault on Bus 8, just as described in Chapter 3. Use your code and Simulink mode from Chapter 3, with the bus and line data, which include the PV system and also running PV_values.m; with the PV model included in your four-machine Simulink model with a three-phase fault from Chapter 3. Make sure the governor is connected and the fault clearing time is reduced from 0.1 to 0.01 s. Further, reduce the overall simulation length to 2 s.

Fig. 7.29 shows the response of the PV generation to a short-term three phase to ground fault at Bus 8, which is cleared after 0.01 s. There are three traces at the top of this figure, which look almost identical until we consult the zoomed view in Fig. 7.30. These three traces show the terminal voltage

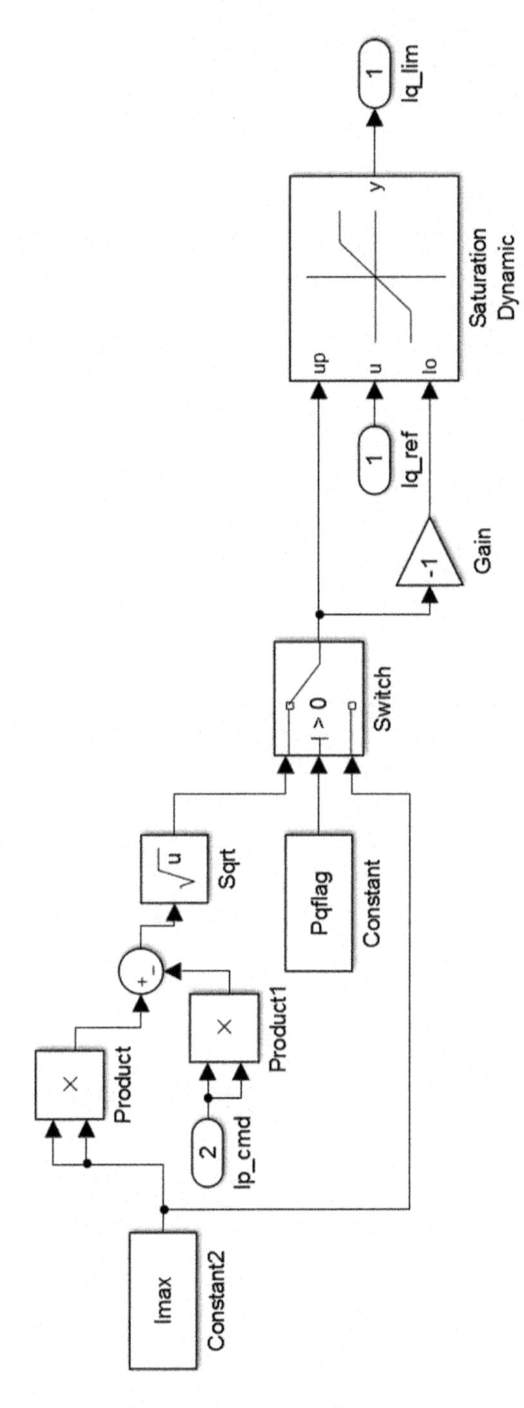

Figure 7.26 Photovoltaic saturation characteristic for Iq_lim.

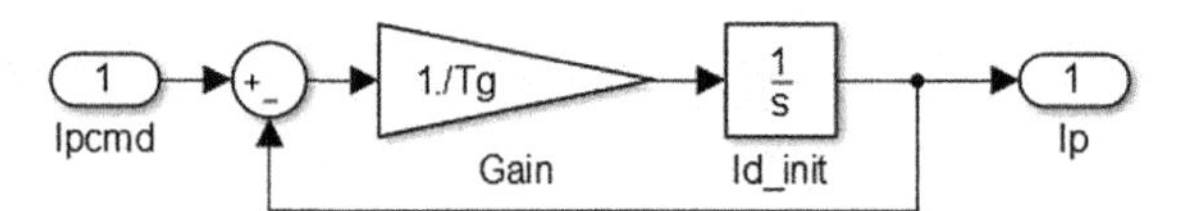

Figure 7.27 Delay block for p-component current.

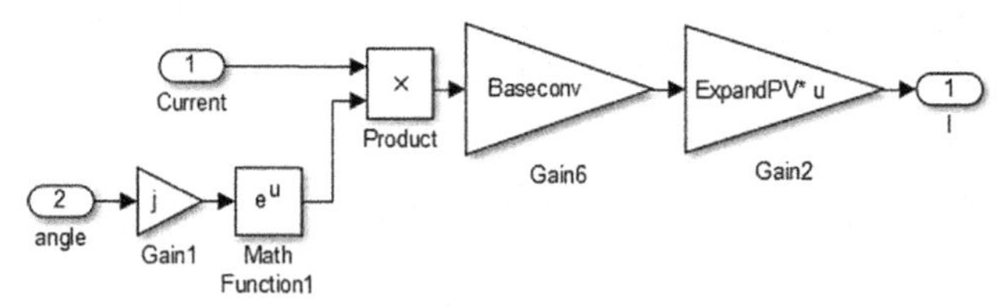

Figure 7.28 Conversion of current to network reference frame, system base and expansion of current vector to number of network buses.

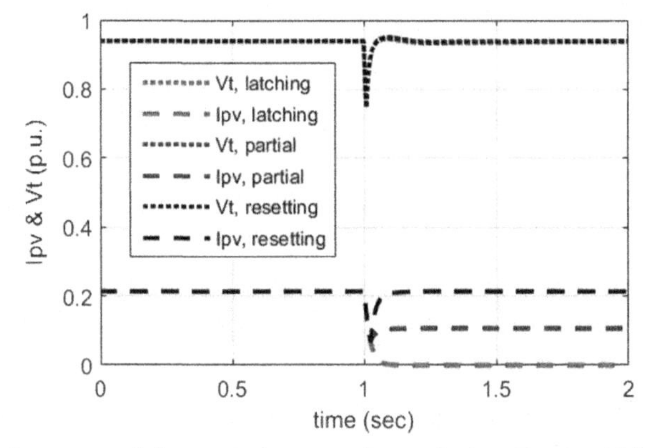

Figure 7.29 Response of photovoltaic generation to fault at Bus 8, which is cleared after 0.01 s.

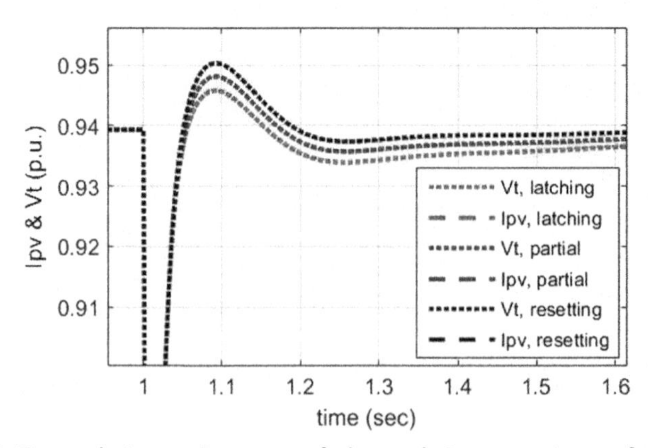

Figure 7.30 Zoomed view – Response of photovoltaic generation to fault at Bus 8, which is cleared after 0.01 s.

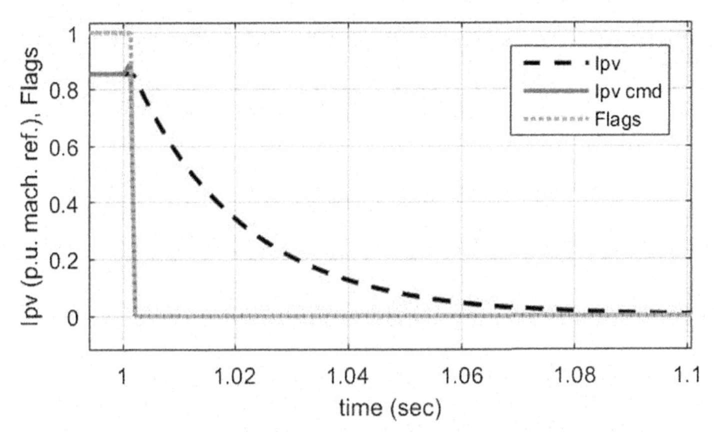

Figure 7.31 Zoomed view – Response of photovoltaic generation to fault at Bus 8, which is cleared after 0.01 s, latching.

at the PV bus (Bus 14). The remaining three traces show the current injections of the PV generation into the network, for three scenarios. The red trace is for PV generators that are set up as latching, meaning the values of tripping flags can only decrease and will not go up and analogously generation is automatically disconnected but never automatically reconnected. The blue trace is for the partially automatic resetting option; in this case, we have chosen 0.5 both for vrrecov and frecov. Finally, the black trace is for a completely automatic resetting option.

To better understand the current response shown in Fig. 7.29, we take a closer look at the steps that lead from a change in the 'flags' setting to a changed output current, for all three cases.

Fig. 7.31 shows the response of the PV generation, when it is set to latching. With the sudden drop in voltage during the fault, the flag value drops almost instantly and does not recover, as this is the latching setting. The command to the PV generation for the current output almost instantly drops to zero as well. The delay block between the command value and the actual current output leads to a drop of the PV generator current to zero after roughly 0.8 s.

In Fig. 7.32, half of the PV generation autoreconnects. As before, the flag value very quickly drops to zero, with the rapid plummeting voltage. This time, as the fault is cleared, the flag value is eventually restored to half the original value and the current command tracks the same behaviour. As seen in Fig. 7.31, the actual current is slower to drop, due to the delay block. At the time the fault has cleared, it has only dipped by a bit below the post-fault reference value, dictated by the current command.

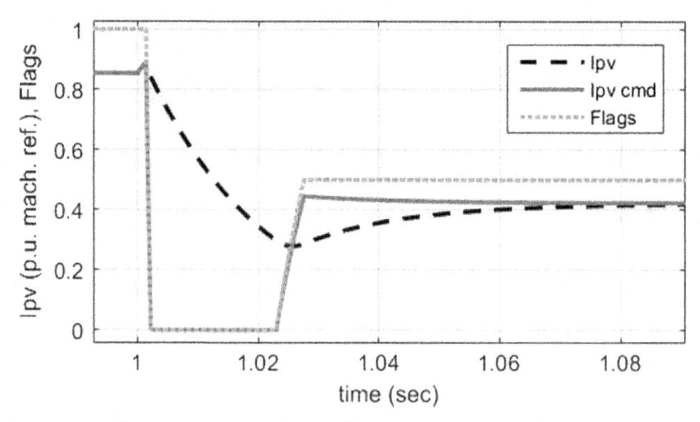

Figure 7.32 Zoomed view – Response of photovoltaic generation to fault at Bus 8, which is cleared after 0.01 s, partial (0.5) autoreconnection.

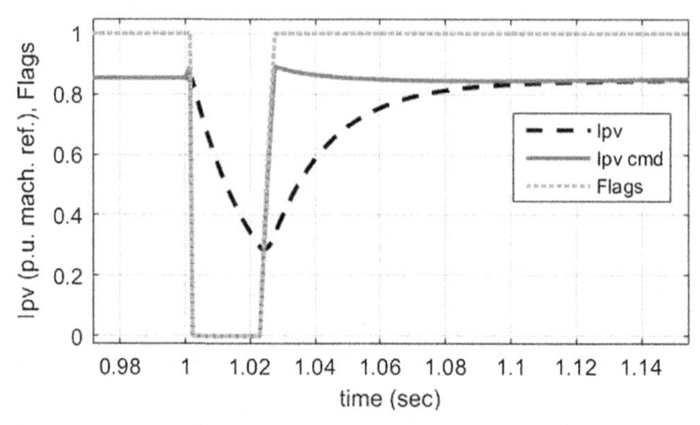

Figure 7.33 Zoomed view – Response of photovoltaic generation to fault at bus 8, which is cleared after 0.01 s, completely autoreconnecting.

Having understood the first two scenarios, the scenario that the PV system is completely autoreconnecting is easy to understand. The fault behaviour is shown in Fig. 7.33. You are suggested to go through it and see if you can explain what is happening here.

References

Ellis, A., n.d. http://www.uwig.org/pvwork/5-ellis-pvsystemmodels.pdf. Retrieved from PV System Models for System Planning and Interconnection Studies.

International Energy Agency, 2019. Snapshot of global PV Markets. Report IEA PVPS T1-35: 2019. http://www.iea-pvps.org/fileadmin/dam/public/report/statistics/IEA-PVPS_T1_35_Snapshot2019-Report.pdf.?.

Lawson, B., n.d. Solar Power. Retrieved from: https://www.mpoweruk.com/solar_power.htm.

Ramachandran, Y., 2005. Modelling of Grid Connected Geographically Dispersed PV Systems for Power Systems Studies. Northumbria University.

Tan, Y.T., 2014. Impact on the power system with a large penetration of photovoltaic generation. Manchester, UK.

WECC, Apr. 2014. WECC Solar Plant Dynamic Modeling Guidelines.

WECC, Mar. 2015. Central Station Photovoltaic Power Plant Model Validation Guideline.

WECC, Sep. 2012. Generic Solar Photovoltaic System Dynamic Simulation Model Specification (Retrieved from Generic solar photovoltaic system dynamic simulation model specification). https://www.google.com/url?sa=t&rct=j&q=&esrc=s&source=web&cd=1&ved=2ahUKEwja_MuynLnkAhUDWBoKHaETCdwQFjAAegQIBRAC&url=https%3A%2F%2Fwww.wecc.biz%2FReliability%2FWECC%2520Solar%2520PV%2520Dynamic%2520Model%2520Specification%2520-%2520September%25202012.pdf&usg=AOvVaw3Dwhi4B4JssA2gMLKJwXY1.

Yashodhan, A., 2014. Control and Operation of Power Distribution System for Optimal Accomodation of PV Generation. London, UK.

CHAPTER EIGHT

Modelling of flexible AC transmission system devices

8.1 Introduction

Since the early days of power systems until now, electricity has been generated predominantly by synchronous machines and transmitted to load centres using AC transmission lines. The power transmission capacity, voltage stability, angle stability and control capability of the grid are limited by physical properties of the network such as resistance, inductance and capacitance. Limited voltage control and reactive power support are achieved through generator excitation system control, transformer tap changing, fixed capacitance or inductance and synchronous condensers. However, the capacity of these devices is not sufficient for the operation of modern power grids. Hence, the capacity of many transmission lines around the world is curtailed to ensure the stability and security of the grid.

Let us consider a scenario where power is transmitted from a remote generator at Bus m to a load centre at Bus-k as shown in Fig. 8.1. The voltages at both the ends of the transmission line are $V_k \angle \theta_k$ and $V_m \angle \theta_m$. Impedance of the transmission line is jX_l. Simplified equations for active power (P_{flow}) and reactive power (Q_{flow}) flowing through the transmission line are given by $P_{\text{flow}} = V_k V_m \sin(\theta_k - \theta_m)/X_L$ and $Q_{\text{flow}} = \left(V_r^2 - V_s V_r \cos\delta\right)/X_L$. As evident from the equations, the active power flow depends on difference in angle and the transmission line impedance. The maximum power transfer capacity of the transmission line, $P_{\max} = V_k V_m / X_L$, can be improved to $P'_{\max} = V_k V_m / (X_L - X_c)$ using capacitor compensation. Flexible AC transmission system (FACTS) devices such as thyristor controlled series compensator (TCSC) can dynamically provide this compensation to improve transmission capacity and enhance system stability.

Unlike active power, the reactive power flow depends on difference in terminal voltages. On high voltage transmission system, it is not sensible to send reactive power over long distance, as it will cause large voltage drop. Hence, adequate reactive power supply should be ensured locally at different parts of a network.

Simulation of Power System with Renewables
ISBN: 978-0-12-811187-1
https://doi.org/10.1016/B978-0-12-811187-1.00008-1
© 2020 Elsevier Inc.
All rights reserved.

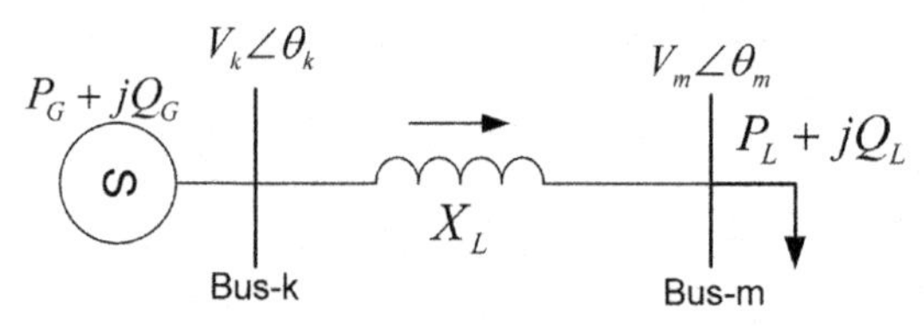

Figure 8.1 A transmission line.

FACTS devices, as the name indicates, offer flexibility on otherwise inflexible network parameters of a transmission line and allow voltage or reactive power injection at a bus. They consist of power electronic switches, inductance and/or capacitances and intelligent controllers and offer fast and precise control of grid parameters. Static VAR compensators (SVCs), TCSCs and unified power flow controllers (UPFCs) are some examples of FACTS devices.

We will briefly discuss some of these devices in the next section. Then more detailed modelling of two devices, SVC, a shunt-connected FACTS device, and TCSC, a series-connected FACTS device, is explained. The chapter helps readers to build a Simulink model of the two devices and provides example results for the readers to make a comparison with their own model. The model developed in this chapter is used to study dynamic stability of a two-area test system and to design a power oscillation damping (POD) controller.

8.2 Flexible AC transmission system devices

In the last few decades, voltage, current and switching frequency of the power electronic converters have improved significantly. This made the development of many FACTS devices possible – some of them are still in the proof-of-concept stage – to provide fast control of the electricity grid. These devices provide fast and accurate current, voltage or impedance control within less than a second. The FACTS devices can be classified according to the way they are connected: shunt-connected, series-connected devices, shunt- and series-connected devices.

SVC is one of the most popular FACTS devices. It is a shunt device, which provides fast reactive power control at the connected bus. STATic COMpensator or STATCOM is another popular device similar to SVC, but it uses a voltage source converter (VSC)–based technology to provide fast reactive power/voltage control. A schematic representation of an SVC and STATCOM is shown in Figs. 8.2 and 8.3, respectively. As indicated

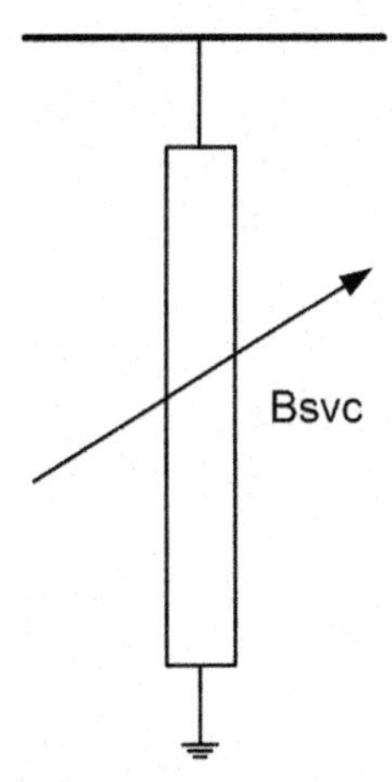

Figure 8.2 Schematic representation of a static VAR compensator (SVC).

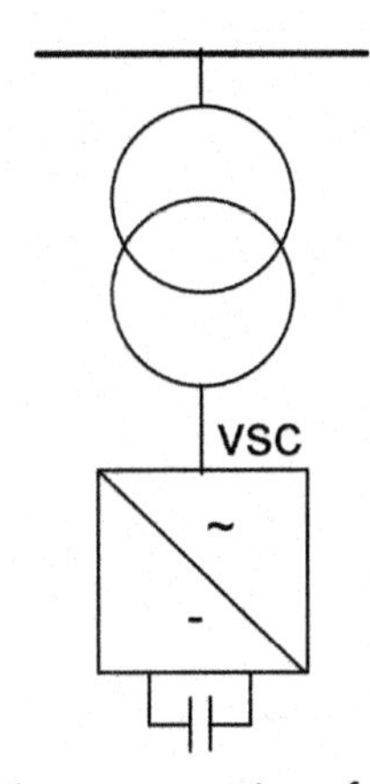

Figure 8.3 Schematic representation of a static compensator.

in the figure, an SVC varies impedance at a bus, whereas STATCOM injects voltage using a VSC to vary reactive power consumption or injection at a bus. By controlling reactive power injection or consumption, the shunt devices can improve power transfer capacity, voltage level and network stability.

Series compensation, a fixed capacitor (FC) in series with a transmission line, used to be installed in 1950s to compensate (reduce) the otherwise constant impedance of the transmission line. This reduces the electrical length of the transmission line and improves power transfer capacity of the transmission line. However, they can cause subsynchronous resonance in certain operating conditions.

A TCSC is a FACTS device alternative for the FC. Fig. 8.4 shows a schematic representation of a TCSC. Apart from providing the same benefit as a fixed series compensator, a TCSC can improve damping of electromechanical oscillations and eliminate possible subsynchronous oscillation. The static synchronous series compensator (SSSC) is another series-connected FACTS device, which uses a VSC to inject a voltage in series with the line as shown in Fig. 8.5. The voltage injected by an SSSC is in quadrature of the current resulting in a variable capacitive or inductive compensation. Another series-connected FACTS device is a short circuit current limiter (SSCL). An SSCL consist of a TCSC and an inductive reactance connected in series as shown in Fig. 8.6. Under normal operating conditions, the effect of inductive

Figure 8.4 Schematic representation of a thyristor controlled series compensator (TCSC).

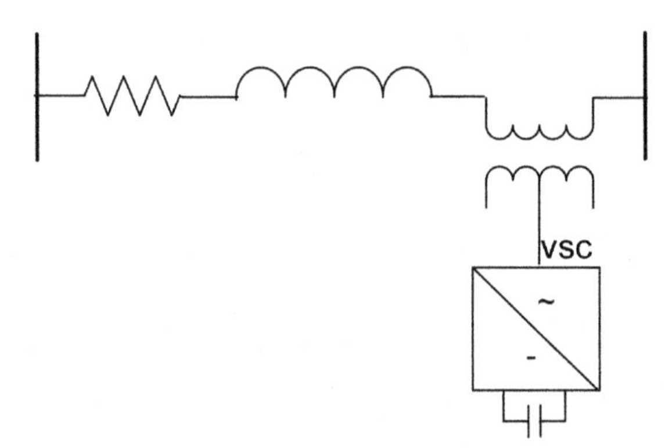

Figure 8.5 Schematic representation of a static synchronous series compensator (SSSC).

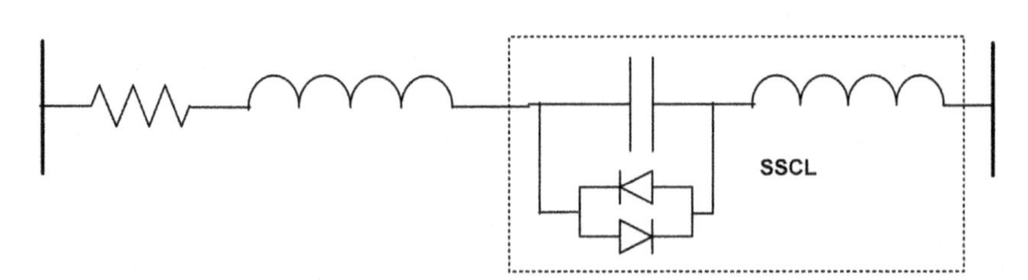

Figure 8.6 Schematic representation of a short circuit current limiter (SSCL).

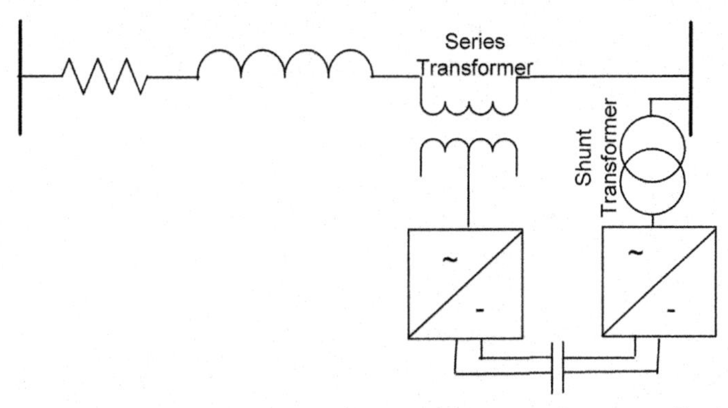

Figure 8.7 Schematic representation of an unified power flow controller (UPFC).

reactance is compensated by capacitive reactance of the TCSC, and during a short circuit, the capacitor is short-circuited so that the inductor limits short circuit current.

The third configuration of FACTS device is made by combining shunt and series compensating devices. A UPFC uses advantages of both shunt compensation and series compensation as shown in Fig. 8.7. Two VSCs using a common capacitor are used to inject power through the series and shunt-connected transformers. This setup supports full control of voltage and power flow.

8.2.1 Applications

Increased power transfer capacity of a transmission line is one of the immediate benefits of using a FACTS controller. A simplified equation for power transfer between two buses is given by $P = V_1 V_2 \sin \delta / X$, where V_1 and V_2 are magnitude of bus voltages, δ is the angle difference of the two bus voltages and X is the reactance of the line connecting the two buses. FACTS devices mentioned above can be used to modify one or more of the above four quantities to increase the power transfer capacity of a line. Similarly, they can also support power flow control. As these devices provide reactive power compensation, the reactive power flow in the network is reduced. This improves voltage in the network, improves stability and reduces losses due to reduced current flow. The fast and reliable control of these devices improves power quality and helps mitigate flicker.

8.2.1.1 Example system using SVC and TCSC

In this chapter, we are going to build Matlab/Simulink programs for SVC and TCSC. Before we discuss their modelling in detail, let us explore the effect of shunt and series compensation using a small test system. Fig. 8.8A–F show the three-bus system used to demonstrate the effectiveness of the two FACTS devices. Electricity generated by a synchronous machine in the Bus 1 is transferred to a load in Bus 3 through a double-circuit transmission line. The top three figures show a case when the system is in a normal operating condition, whereas the bottom three figures show the state of the system when one of the transmission lines is tripped.

Fig. 8.8A,B show the cases without using FACTS devices. The voltage at Bus 3 in the normal operating condition is 0.98 pu, whereas following the loss of a transmission line, the voltage drops to 0.91 pu An SVC Bus 3 or a TCSC at one of the transmission lines can improve the voltage level at Bus 3 following the contingency.

Fig. 8.8C,D show the case when an SVC is installed at Bus 3 to improve the voltage level. With 0.62 pu reactive power from the SVC, the prefault at Bus 3 is 1 pu If one of the transmission lines trips, the voltage at the Bus 3 reduces. However, the SVC is programmed to control Bus 2 voltage, and it changes its output to 1.3 pu and maintains Bus 3 voltage at prefault level of 1 pu.

Fig. 8.8E,F demonstrate the use of TCSC in the system. With a 10% compensation for the series reactance, prefault voltage at bus 3 is 0.981 pu. When one of the transmission lines trips, to maintain system stability and voltage levels, the TCSC compensation ratio is increased to 50%. Now the voltage at Bus 3 is 0.966 pu. Although the voltage did not recover to prefault level, it is much better than the case when there is no TCSC installed.

8.3 Static VAR Compensator

A SVC is a shunt-connected device, which can generate and consume reactive power. This makes it an ideal device to control network voltage. It consists of one or more thyristor controlled reactors (TCRs) and thyristor switched capacitors (TSCs). A transformer is generally used to interface them with the high voltage bus. Filters are used to remove harmonic currents, and at fundamental frequency, they are capacitive. Often, an SVC is part of a static compensation system. Static VAR system (SVS) consists of additional FCs, mechanically switched capacitors (MSC) and saturated reactors. Fig. 8.9 shows a schematic diagram of an SVS IEEE, 1994. It consists of a TSR, two TSCs and an MSC. The capacity of TCR is larger than the discrete TSC and MSC block to facilitate a smooth control over the entire SVS rating.

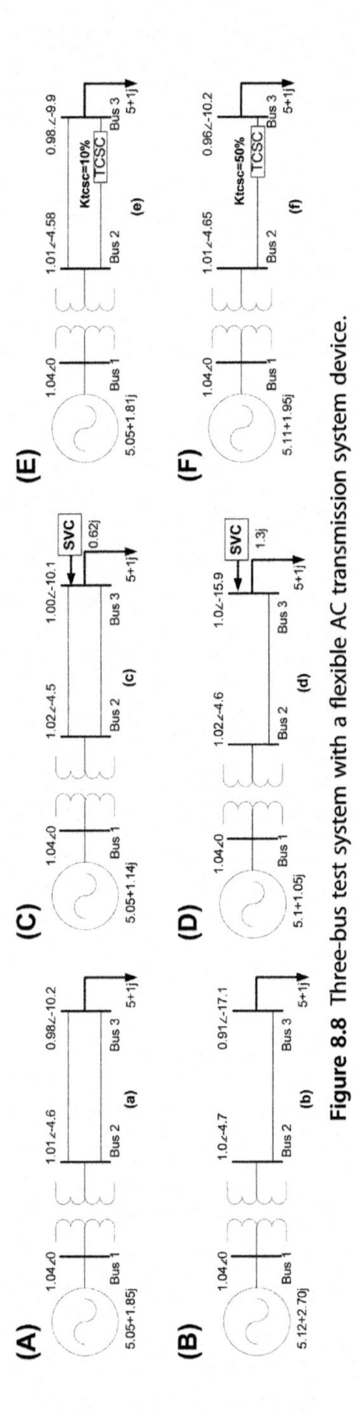

Figure 8.8 Three-bus test system with a flexible AC transmission system device.

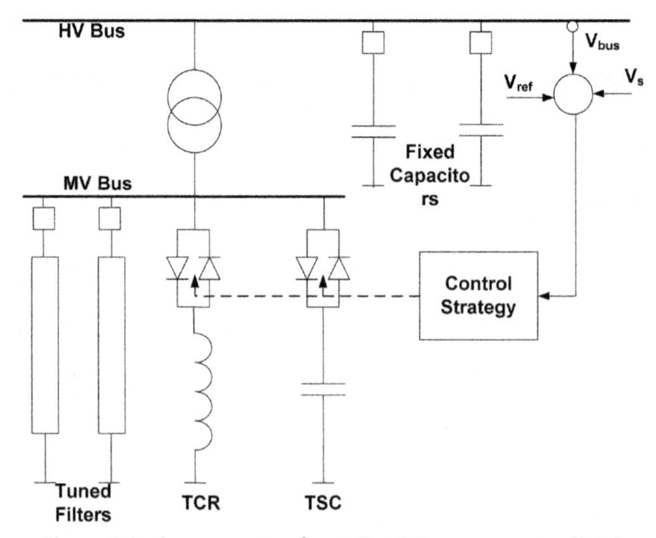

Figure 8.9 Components of a static VAR compensator (SVC).

By continuously varying the inductive reactance, the effective susceptance or net reactive power injection at a bus is controlled. The net susceptance of SVC is given by $B_{svc} = B_C - B_L$, where B_C is susceptance of capacitor and B_L is the susceptance of inductor. The reactive power injection at a bus is given by $Q_{svc} = V_{svc}^2 B_{svc}$.

Steady-state response of the SVS is shown in Fig. 8.10. In the active control range, current/susceptance and reactive power are varied to regulate voltage according to SVS droop (slope) characteristics. The slope is selected

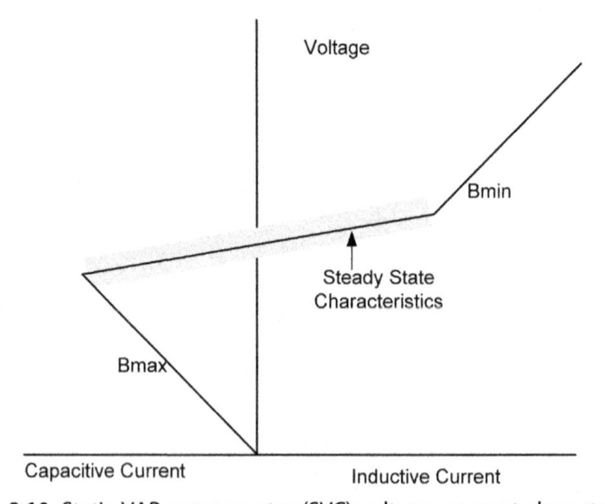

Figure 8.10 Static VAR compensator (SVC) voltage–current characteristics.

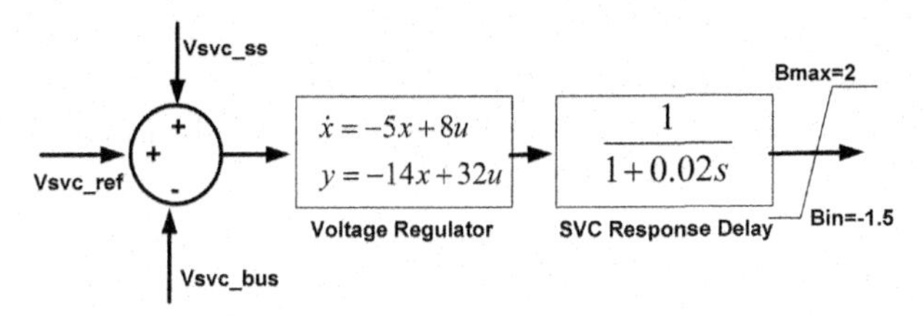

Figure 8.11 Dynamic model of a static VAR compensator (SVC).

by considering the characteristics of other reactive power/voltage control sources in the power system. The value is typically around 1%–5%. Hingorani and Gyugyi (1999). A supplementary control can be added to the SVC to modulate the bus voltage and thereby damping electromechanical oscillations in the network. Such a controller is called POD controller. In Chapter 9, the design of a POD controller using SVC is demonstrated. Placing SVCs in strategic locations in the grid can improve power transfer capacity, voltage level and network stability.

8.3.1 Modelling of static VAR compensator

The dynamic models of SVC described in literature vary slightly. Overall, they represent the control characteristics of the SVC system. For modelling and stability studies discussed in this book, we will use the model of SVC shown in Fig. 8.11. We will assume that the SVC consists of a TCR of 150 MVAr capacity and two TSCs of 100 MVAr capacity each. This provides a control range of −150 MVAr–200 MVAr, which corresponds to a susceptance range of −1.5 pu–2 pu at 1 pu voltage. The model consists of a voltage regulator and an SVC response delay transfer function. Inputs to the block are an SVC bus voltage (Vsvc_bus), SVC bus voltage reference (Vsvc_ref) and a supplementary input (Vsvc_ss). The supplementary input can be used to design the POD controller. We will demonstrate the POD design using SVC in Chapter 9.

8.4 Thyristor controlled series compensation

A TCSC is formed of a TCR connected in parallel with a FC to provide rapid and continuous variation of transmission line series reactance. A variable capacitance effect is created by controlling the thyristor valves. Like SVC, TCSC is also a well-established technology with number of installations around the world.

A TCSC primarily modifies the line impedance of a transmission line and thereby helps to control the power transfer capacity of the transmission line. It is also widely used to damp electromechanical oscillations. The control algorithm can remove potential subsynchronous oscillation issues present in fixed series compensation. In Chapter 9, we will demonstrate the use of TCSC control to improve damping of electromechanical oscillations.

8.4.1 Modelling of thyristor controlled series compensator

Fig. 8.12A shows a schematic representation of a transmission line with a TCSC, in which the variable capacitor represents the TCSC [Pal and Chaudhuri]. The compensated impedance of the transmission line is given by $Z_l = R_l + jX_l(1 - k_c)$, where k_c is the TCSC compensation ratio given by $k_c = X_c/X_l$. In the dynamic modelling of the system, a TCSC can be represented using a current injection model. Instead of the variable capacitor, two variable current injections at the terminal buses represent the effect of the TCSC. The transformation of the TCSC model from a variable series capacitor to injected current at terminal buses is depicted in Fig. 8.12A,D. Let I_l be the current flowing from Bus m to Bus k. Voltage drop across TCSC is $V_{\mathrm{tcsc}} = -1X_cI_l$. The variable capacitor can be replaced by a variable voltage source as shown in Fig.8.10B. The voltage source V_{tcsc} can be

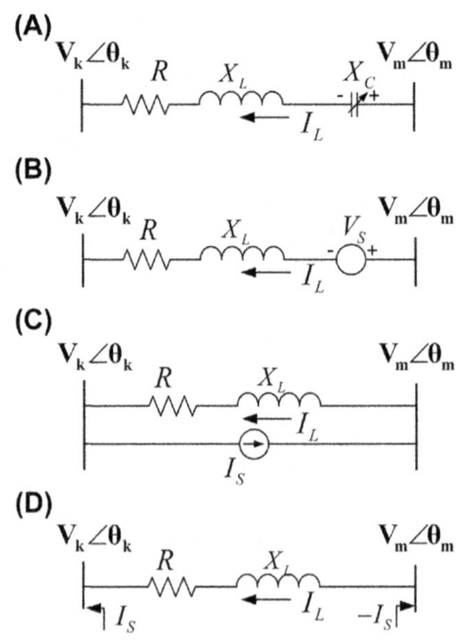

Figure 8.12 Different power flow representations of a thyristor controlled series compensator (TCSC).

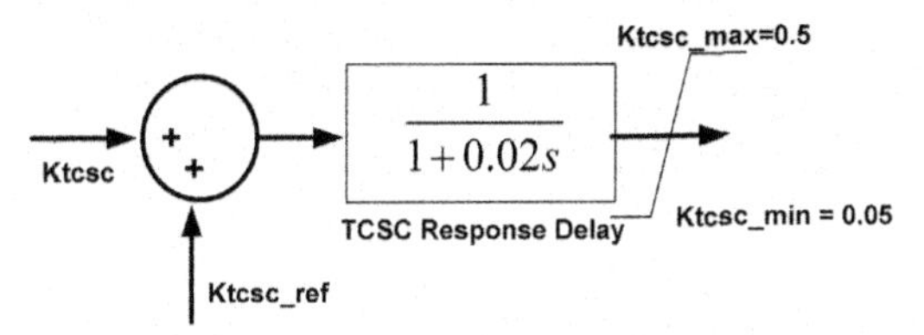

Figure 8.13 Dynamic model of a thyristor controlled series compensator (TCSC).

replaced by a current source $I_{tcsc} = V_{tcsc}/Z_l$ as in Fig. 8.12C. Furthermore, the current source I_{tcsc} is split into two current injections at terminal buses k and m as shown in Fig. 8.12D.

For small signal stability studies, the dynamic characteristics of TCSC is usually modelled with a single time constant T_{tcsc} as shown in Fig. 8.13. While T_{tcsc} is the response time of the TCSC, k_{tcsc} is the reference setting of the TCSC and $k_{tcsc-ref}$ is the supplementary signal for the damping control. Modulating the supplementary signal using a POD is used to improve the damping of electromechanical oscillations in the power system. We will discuss the design of a POD using TCSC in Chapter 9.

8.5 Implementation of SVC and TCSC models

In the previous section, we have discussed a simple dynamic model of SVC and TCSC. Let us implement this on a three-bus system powered by a synchronous generator. The power flow program, model of synchronous generator and the initialization program developed in Chapter 2 can be used for the system. Only the bus and line matrices require modification.

8.5.1 Power flow solution considering SVC and TCSC

Let us simulate the three-bus system example where a line fault occurs at 1 s. This will cause a drop in Bus 3 voltage. The SVC voltage controller will regulate the Bus 3 voltage by changing its reactive power output. For the TCSC, the K_{tcsc} input is varied to simulate its effect.

The dynamic simulation of the system starts when the power system is operating under the normal operating condition. After 1 s, one of the transmission lines trips changing the network configuration. To simulate both states of the system, the simulation model needs two impedance matrices representing the prefault and postfault operating condition. During the simulation, a switch can be used to change the matrix used for the calculation at the specified time. So, before the start of the simulation, we must calculate a prefault and a postfault impedance matrix.

Secondly, for initializing this simulation, the power flow solution considering the steady-state model of SVC and TCSC at the normal operating condition is required. However, a simple power flow program is discussed in Chapter 1, and we would like to use it throughout this book. With a small workaround, the simulation explained in this chapter can be performed using the program given in Chapter 1. Readers are advised to use an advanced power flow program for simulating more complicated networks.

Changes in power flow program for SVC: In the case of an SVC, its reactive power output can be specified as reactive power generation at the bus in the bus matrix. This will not affect the calculation of the prefault and postfault impedance matrix and will not affect any other device connected. Script 8.1 and Script 8.2 show the required program to find the power flow solution as well as the prefault impedance and postfault impedance for the three-bus system using SVC.

Changes in power flow program for TCSC: In the case of a TCSC, modify the line matrix. Reduce the series reactance of the transmission line in the line matrix to reflect the effect of TCSC while computing the power flow solution. However, while calculating the impedance matrices, the actual series impedance of the line without considering the TCSC must

```
clear all

system_base_mva = 100.0;
s_f=60; %System frequency in Hz
s_wb=2*pi*s_f; %Base value radial frequency in rad/sec
s_ws=1; %p.u. value of synchronous speed


% bus data without SVC or TCSC
bus = [...
1 1.04    00.0    0.00     0.00 0.00    0.00     0.00   0.00 1;
2 1.00    00.0    0.00     0.00 0.00    0.00     0.00   0.00 3;
3 1.00    00.0    0.00     0.00 5.00    1.00     0.00   0.00 3;
];

% Line data without SVC or TCSC
line = [01 02 0.00 0.0167 0.00 1.0 0.0;
        02 03 0.004 0.040 0.07 1.0 0.0;
        02 03 0.004 0.040 0.07 1.0 0.0];
% *********************** MACHINE DATA  ***********************
mac_con =[
1 900 0.2 0.0025 1.8 0.3 0.25 8 0.03 1.7 0.55 0.25 0.4 0.05 6.5 0];

% *********************** EXCITATION SYSTEM DATA ******************
s_Ka=200;%Static excitation gain Padiyar p.328
s_Ta=0.02;%Static excitation time constant Padiyar p.328

% **** Governor Control SYSTEM DATA  *********
s_Tg=0.2;%Kundur p.598
s_Rgov=0.05;%Kundur p.598
```

Script 8.1 Bus and line matrices for the three-bus system.

```
% execute after Script 8.1

% changes for SVC
svc_bus = 3;      % specify svc bus
bus(svc_bus,5) = 0.62;  % specify reactive power output of SVC

% calculate pre-fault admittance matrix
[Y] = form_Ymatrix(bus,line);
% calculate pre-fault power flow solution
[bus_sol, line_flow] = power_flow(Y,bus, line);

% Add the loads to Y matrix, this is the pre-fault case
YPL = bus_sol(:,6)./bus_sol(:,2).^2;
YQL = bus_sol(:,7)./bus_sol(:,2).^2;
Y = Y + diag(YPL-1i*YQL);

% find pre-fault impedance
Z = inv(Y);

% change line matrix to reflect tripping of one of the lines
line(2,:) = [];
% recalculate Y matrix, post fault
[Yf] = form_Ymatrix(bus,line);
% Add the loads to Y matrix, this is the post-fault case
Yf = Yf + diag(YPL-1i*YQL);

% find post-fault impedance
Zf = inv(Yf);

% Synchronous machine initialisation

Smachs=[1];% buses with synchronous machines

Nbus=size(bus,1); % Number of buses
NSmachs=size(Smachs,1); % Number of synchronous machines

% program to initialise the synchronous machines

Synch_parameter_sys_base % or _gen_base, depending on the base system we
%decide to work on.

generic_Sync_Init

% program to initialise svc

svc_init
```

Script 8.2 Script to find power flow, prefault impedance and postfault admittance for the three-bus system with static VAR compensator (SVC).

be used. This is because the TCSC is modelled separately as power injection in the dynamic simulation. Script 8.1 and Script 8.3 show the programs to find the power flow solution, prefault impedance and postfault impedance for the three-bus system using TCSC.

8.5.1.1 Representation of static VAR compensator

The model of an SVC shown in Fig. 8.9 can be programmed in Simulink as shown in Figs. 8.14 and 8.15. The output of the generator block and SVC

```
% excecute after Script 8.1

% changes for TCSC
tcsc_k = 2;  % tcsc from bus
tcsc_m = 3;  % tcsc to bus
tcsc_indx = 3; % index of the line in the Line matrix where tcsc is
%installed
tcsc_KC = 0.1;  % tcsc compensation ratio : xc/xl
tcsc_Ttcsc = 0.02;  % time constant for TCSC model

% impedance of the transmission line without TCSC
Z_TCSC = line(tcsc_indx,3)+1i*line(tcsc_indx,4);
XL_TCSC = 1i*imag(Z_TCSC);
RL_TCSC = real(Z_TCSC);

Nbus = size(bus,1);
ExpandTCSC = zeros(Nbus,1);
ExpandTCSC(tcsc_k) = -1;
ExpandTCSC(tcsc_m) = 1;


% modify series reactance of transmission line to reflect presence of
%TCSC
line(tcsc_indx,4) = line(tcsc_indx,4)*(1-tcsc_KC);

% Obtain load flow solution with TCSC
[Y] = form_Ymatrix(bus,line);
% calculate pre-fault power flow solution
[bus_sol, line_flow] = power_flow(Y,bus, line);

% find steady state voltage at the tcsc buses - required for the simulink
%program
tcsc_vk = bus_sol(tcsc_k,2)*exp(1i*bus_sol(tcsc_k,3)*pi/180);
tcsc_vm = bus_sol(tcsc_m,2)*exp(1i*bus_sol(tcsc_m,3)*pi/180);

% Obtain impedance of load
YPL = bus_sol(:,6)./bus_sol(:,2).^2;
YQL = bus_sol(:,7)./bus_sol(:,2).^2;

% Remove the compensation from the transmission line
line(tcsc_indx,4) = line(tcsc_indx,4)/(1-tcsc_KC);

% find pre-fault impedance
[Y] = form_Ymatrix(bus,line);
Y = Y + diag(YPL-1i*YQL);
Z = inv(Y);

% apply fault - drop a line
line(2,:) = [];

% find post-fault impedance
[Yf] = form_Ymatrix(bus,line);
Yf = Yf + diag(YPL-1i*YQL);
Zf = inv(Yf);


% Synchronous machine initialisation

Smachs=[1];% buses with synchronous machines

Nbus=size(bus,1); % Number of buses
NSmachs=size(Smachs,1); % Number of synchronous machines

% program to initialise the synchronous machines

Synch_parameter_sys_base % or _gen_base, depending on the base system we
%decide to work on.

generic_Sync_Init
```

Script 8.3 Script to find power flow, prefault impedance and postfault admittance for the three-bus system with thyristor controlled series compensator (TCSC).

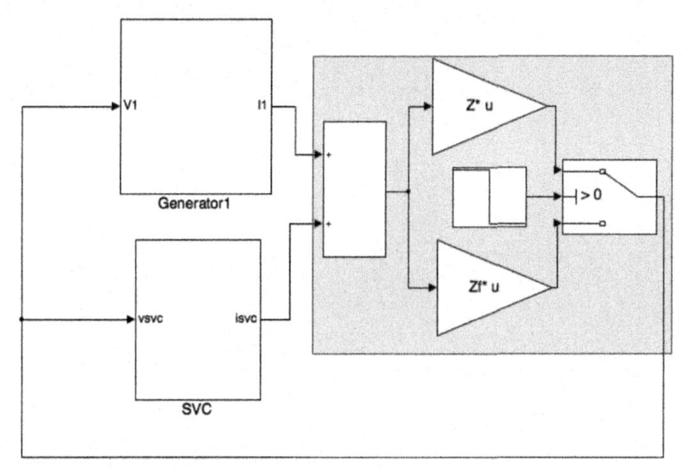

Figure 8.14 Simulink block for the three-bus system with a static VAR compensator (SVC).

block are a vector of three elements. As the SVC is located at Bus 3, the third element of this vector is the current injection at Bus 3 from the SVC. The sum of the generator block output and SVC block output gives the vector of current injection at all buses. We have two matrices Z and Zf representing prefault and postfault operating conditions, respectively. The operating condition is selected using a switch, which selects the first input (prefault) when the second input is more than one and third input (postfault) when the second input is less than or equal to zero. The switch is operated using a step signal that changes its output from one to zero at time 1 s. So, for the first and second input, the calculation uses the prefault impedance, and for the later, the calculation is performed using the postfault impedance.

The internal diagram of the SVC block is shown in Fig. 8.15. The output of the SVC model in Fig. 8.9 is Bsvc, while we would like to obtain corresponding current injection at the bus. This is a simple task of multiplying the complex susceptance and bus voltage. However, a direct feedback of voltage creates an algebraic loop, which Matlab struggles to solve. There are many methods to break an algebraic loop, one of them is to use an initial condition (IC) block. IC block takes initial value of the voltage as input and helps the simulation to start running. The IC block only accepts real values, so the complex voltage magnitude is split into real and imaginary and later changed to imaginary value. This part of the diagram is highlighted using a shaded area in Fig. 8.15. A detailed discussion on the algebraic loop is included in Chapter 5.

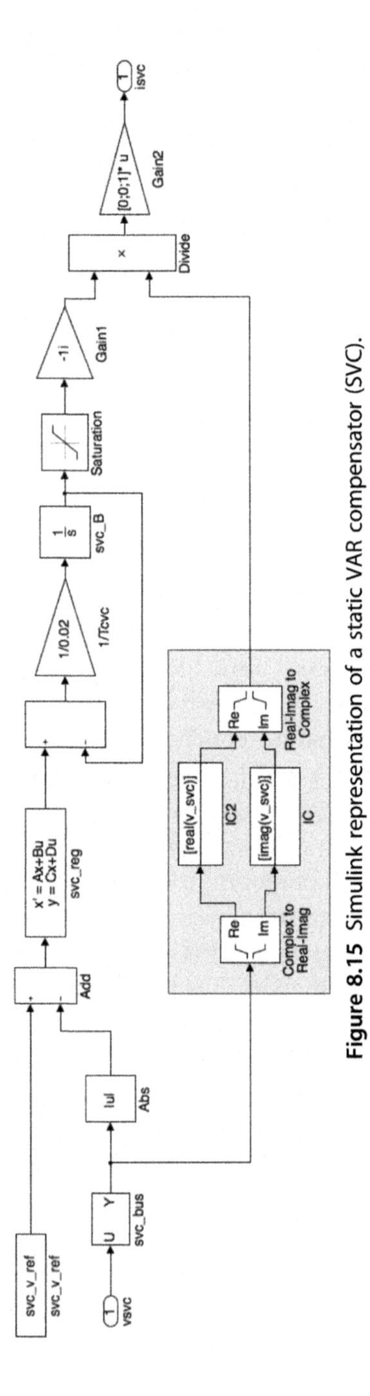

Figure 8.15 Simulink representation of a static VAR compensator (SVC).

```
% find voltage at svc bus
v_svc = bus_sol(svc_bus,2)*exp(1i*bus_sol(svc_bus,3)*pi/180);
% find susceptance of svc from the reactive power specified
B_svc = bus_sol(svc_bus,5)/abs(v_svc)^2;

% specify the svc_regulator transfer function in state space
% format [A B; C D]
svc_reg_ABCD = [-5 8; -14 32];
% solve equation
% |0      |=| A B||[state|
% |output|  | C D||input |
temp = svc_reg_ABCD\[0;B_svc];
% extract the initial value of state for initial value of svc_reg block
svc reg x0 = temp(1);
% extract input constant in the svc model
svc_v_ref = temp(2)+abs(v_svc);

% Matrix to insert svc current at generator buses,
ExpandSVC = zeros(Nbus,1);  % assuming only one SVC
ExpandSVC(svc_bus) = 1;
```

Script 8.4 Initialization program for static VAR compensator (SVC) model.

Like synchronous machines, the SVC block also requires initial values to be specified. Values such as Bsvc, svc_v_ref block initial state and svc bus voltage are calculated using the Script 8.4. Follow instructions on Scripts 8.1, 8.2 and 8.4 to initialize the three-bus system with an SVC. Run the Simulink model, which simulates SMIB system response following a line fault at time 1 s. Compare the results with Fig. 8.16 that shows bus voltages at Bus 1 and Bus 3 and SVC susceptance. The dotted lines show the case if a fixed compensation is used instead of an SVC.

8.5.1.2 Representation of thyristor controlled series compensator

Fig. 8.17 shows the Simulink implementation of the SMIB system with a TCSC. This is like the SMIB with SVC model, except that the SVC block is replaced with a TCSC block. Fig. 8.17 and Fig. 8.18 shows the internal block diagram of the TCSC block that calculates the current injection required in the end buses to simulate the effect of a TCSC. The need for the IC and IC2 block in the voltage loop is explained in the previous section.

To run the simulation, first run Scripts 8.1 and 8.3 following the instructions on them. The Simulink program simulates a line fault at time 1 s. At the same time, the TCSC compensation ratio is changed to 0.5 using a step signal. Example results are shown in Fig. 8.19. This model is used in Chapter 9 to design a POD controller for a two-area test system model.

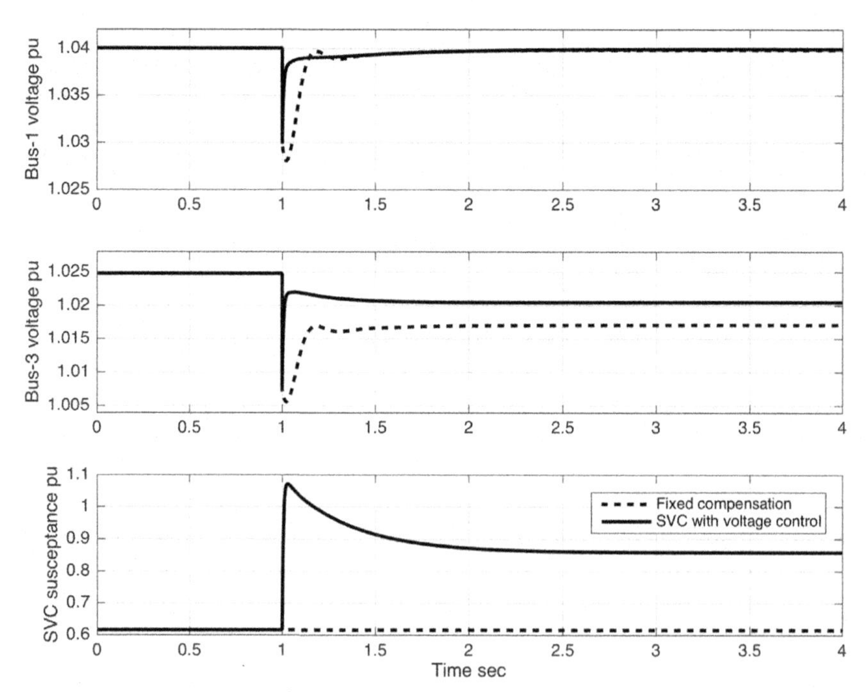

Figure 8.16 Dynamic simulation results of the three-bus system with a static VAR compensator (SVC).

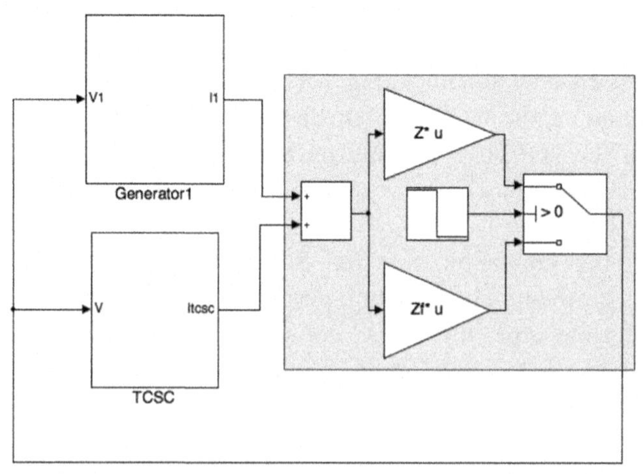

Figure 8.17 Simulink block for the three-bus system with a thyristor controlled series compensator (TCSC).

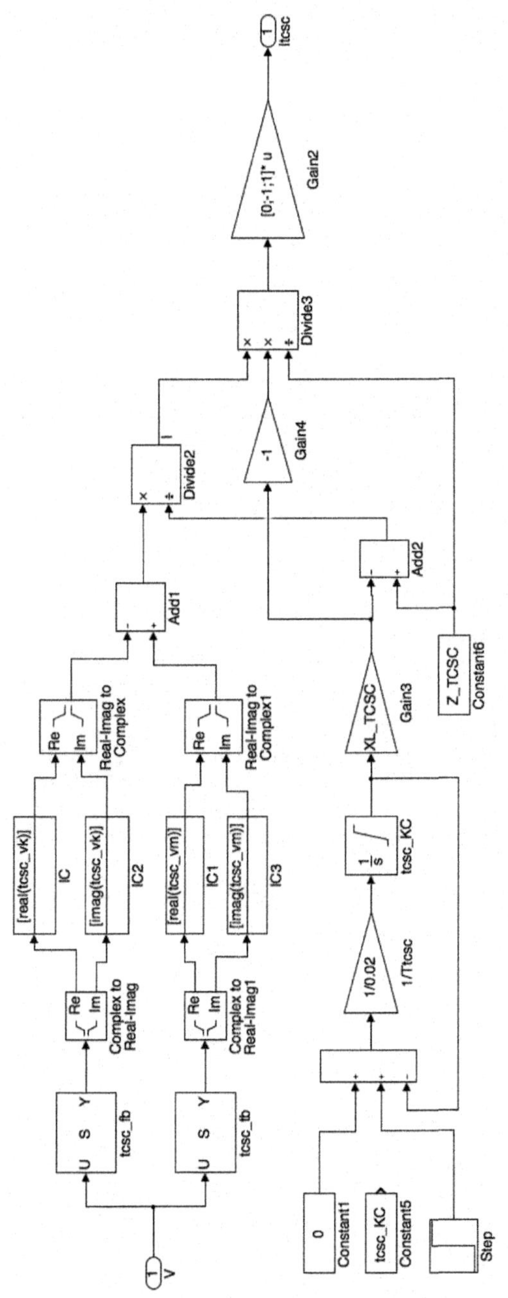

Figure 8.18 Simulink representation of a thyristor controlled series compensator (TCSC).

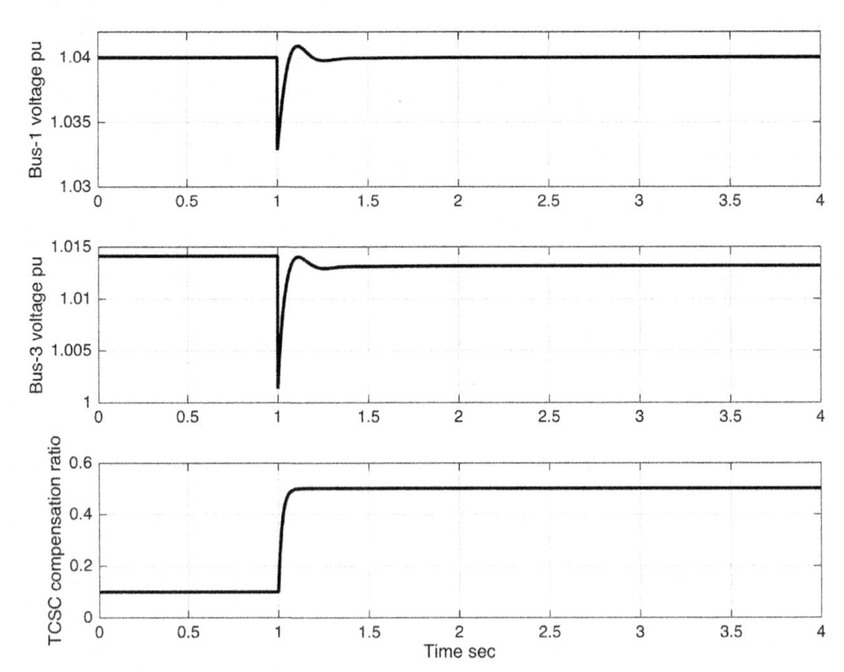

Figure 8.19 Dynamic simulation results of the three-bus system with a thyristor controlled series compensator (TCSC).

References

Hingorani, N.G., Gyugyi, L., December 1999. Understanding FACTS: Concepts and Technology of Flexible AC Transmission Systems. Wiley-IEEE Press.

IEEE Special Stability Controls Working Group, Feb. 1994. "Static VAr compensator models for power flow and dynamic performance simulation," in. IEEE Transactions on Power Systems 9 (1), 229–240.

Pal, B., Chaudhuri, B., Robust Control in Power Systems, Springer US.

CHAPTER NINE

Case study of interarea oscillations in power system

9.1 Introduction

In the previous chapters we have developed the models of a synchronous machine, wind turbine, static VAR compensator (SVC) and thyristor controlled series compensator (TCSC) and tested them individually on a single machine infinite bus (SMIB) test system. Additionally, a four-machine simulation program is developed in Chapter 3. We will extend this program with wind turbines, SVC and TCSC and discuss interarea oscillations and power oscillation damping (POD) controller design. A two-area test system shown in Fig. 2.5 is used for this study. However, the simulation program developed here can be adopted for other networks.

The dynamic stability of the four-machine system is discussed in Section 9.2. Participation factor analysis (Kundur, 1994, Rogers, 2000) is presented, revealing the characteristics of two local modes and an interarea mode present in the system. In Section 9.3 and Section 9.4, we will use TCSC and SVC, respectively, to design a POD to improve the damping of the interarea oscillation mode. For this study, an SVC is installed at Bus 7, or a TCSC is installed in one of the transmission lines connecting Bus 8 and Bus 9. In Section 9.5 and Section 9.6, the synchronous generator at Bus 4 is replaced with a wind turbine generator to study the dynamic interaction of doubly fed induction generator (DFIG) and permanent magnet synchronous generator (PMSG), respectively, to changes in the network. The mac_con matrix should be modified for this case by removing the row corresponding to the synchronous generator at G4.

9.2 Analysis of two-area system

Let us start by testing the two-area system model developed in Section 3.8. Note that the example results shown in this chapter are generated assuming that governors are not connected as discussed in Section 3.9.1. Run the program to simulate a three-phase fault as discussed. You will get similar responses as obtained in Figs. 3.36–3.39. The system

Simulation of Power System with Renewables
ISBN: 978-0-12-811187-1
https://doi.org/10.1016/B978-0-12-811187-1.00009-3
© 2020 Elsevier Inc.
All rights reserved.

shows an undamped oscillation. Notice the changes in active power output of G1 and G3. When G1 is maximum, G3 is minimum and vice versa. It looks like generators at two areas are oscillating together. Let us plot the speed of the generators, which is shown in Fig. 9.1. It is clear that when speed of generators G1 and G2 at area 1 is increasing while speed of generators G3 and G4 at area 2 is reducing and vice versa. Following the three-phase fault, an oscillation is triggered in the system in which generator from different parts or areas in the system participate. Hence, it is called interarea oscillations. What is the frequency of oscillations? Between time equal 2 and 9 s, speed of generator 3 (G3) has little more than four cycles of oscillations, which means time period of oscillation is approximately $T = 7/4$ s. Hence, frequency of oscillation is approximately 4/7 Hz, which is 0.57 Hz.

We can also study stability of the system by obtaining a linearized model of the system. In Chapter 4, we have introduced the commands linmod and eig to obtain a linearized model of a nonlinear Simulink model and eigenvalues of a state matrix, respectively. Try using these commands. Complex eigenvalues of the system at the operating condition presented in Chapter 2 are listed in Table 9.1. Are you getting similar values? Compare frequency of oscillation of seventh eigenvalue with the one we estimated in the last

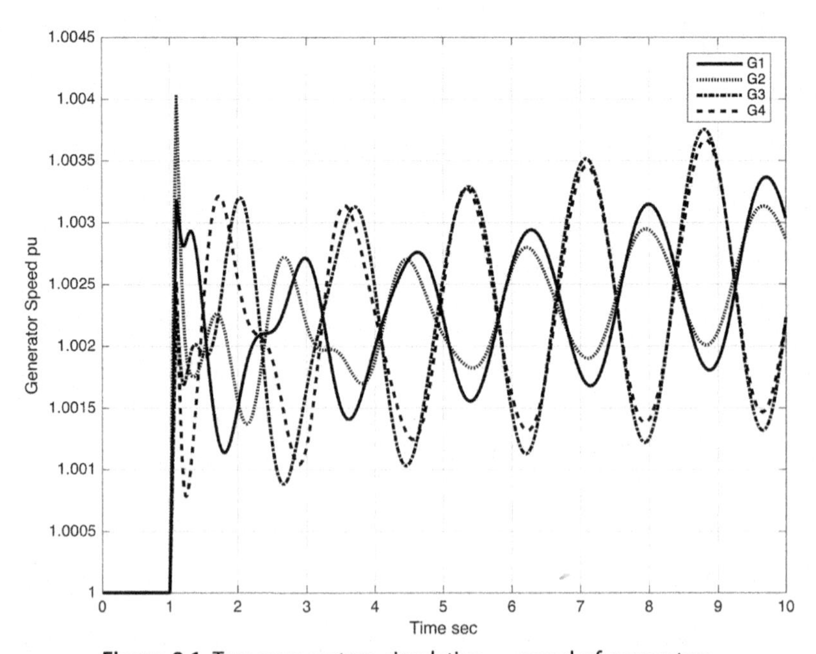

Figure 9.1 Two-area system simulation – speed of generators.

Table 9.1 Complex eigenvalues of the two-area test system.

Index	Eigenvalue	Frequency (Hz)	Damping ratio (%)
1	$-13.6503 + 19.4922i$	3.1023	57.3627
2	$-14.7148 + 16.7249i$	2.6619	66.0548
3	$-17.3557 + 5.6146i$	0.8936	95.1453
4	$-17.4152 + 4.9938i$	0.7948	96.1261
5	$-0.5678 + 7.0682i$	1.1249	8.0072
6	$-0.5828 + 7.1239i$	1.1338	8.1530
7	$0.0362 + 3.6547i$	0.5817	-0.9915

paragraph. The seventh eigenvalue relates to interarea mode. Let us verify this using participation factor analysis.

9.2.1 Participation factor analysis

Use Script 9.1 to find the participation factor corresponding to the 25th eigenvalue. It will be evident that the mode has participation from the rotor angle and rotor speed states of all the synchronous machines. As synchronous machines from two different areas participate in this mode, it is called inter-area mode. Now check the participation factors for the other modes such as local modes.

```
% obtain state space representation

sys = linmod('two_area_synch');
SN = sys.Statename;


% find eigenvalues and eigenvectors
[vright,d] = eig(sys.a);
dr = diag(d);

% form left eigenvector
vleft = inv(vright);
vleft = transpose(vleft);

% specify index of the eigenvalue to perform pf analsysis
index = 25;

% find the dot product of two eigenvectors
dotproduct = abs(vright(:,index)).*abs(vleft(:,index));
% normalise with sum of the product
pf = dotproduct/sum(dotproduct);

strcat(num2str(pf),'-------------------',SN)
```

Script 9.1 Program to find participation factor.

Note: To correctly print the state names, the name of the integration blocks must be modified in the Simulink file.

9.3 Two-area system with a thyristor controlled series compensator

The two-area test system simulated in Section 9.1 has an unstable interarea mode, which limits the transmission capacity of the line. One of the solutions to this problem is to install a TCSC in one of the transmission lines connecting the areas and design a supplementary POD controller. In this section, we will develop a Simulink model of the two-area test system with a TCSC and demonstrate the operation of a TCSC-POD controller.

A POD controller is a feedback controller that modulates the TCSC reference input. Selection of feedback signal for the controller is very important to get the best damping at least control cost. Generally, signals local to the actuator (TCSC in this case) are used; however, with increased interest in wide-area control and monitoring, global signals (signals from other parts of the network) are explored. One of the criteria for selecting the signal is the magnitude of residue for the transfer function between signal and actuator, corresponding to the oscillatory mode with poor damping. This topic is covered widely in the literature, and readers are advised to further refer to this. In Chapter 4, we have discussed the method to calculate residue. For this demonstration, we will show how a signal is selected for the damping controller used in the two-area network.

9.3.1 Simulink model

A Simulink model of the two-area test system with a TCSC is shown in Fig. 9.2. The TCSC model representation consisting of Constant5 block, 1/Ttcsc, tcsc_ref and tcsc_KC is brought out of the TCSC block for clarity of explanation. Compare this figure with the TCSC model in Chapter 8.

The measurement area at the top left-hand side of Fig. 9.2 is used to calculate active power, reactive power and magnitude of line current through a few selected transmission lines. For simplicity, we have neglected the capacitor current in the transmission line in this calculation. Suppose Y_{ab} is the series admittance between Bus a and Bus b and V_a and V_b are the voltages at these buses. Then the current flowing through the line is $I_{ab} = (V_a - V_b)Y_{ab}$. For each transmission line selected, we get three signals: active power real $\left(V_a I_a^*\right)$, reactive power imag $\left(V_a I_a^*\right)$ and current abs (I_{ab}).

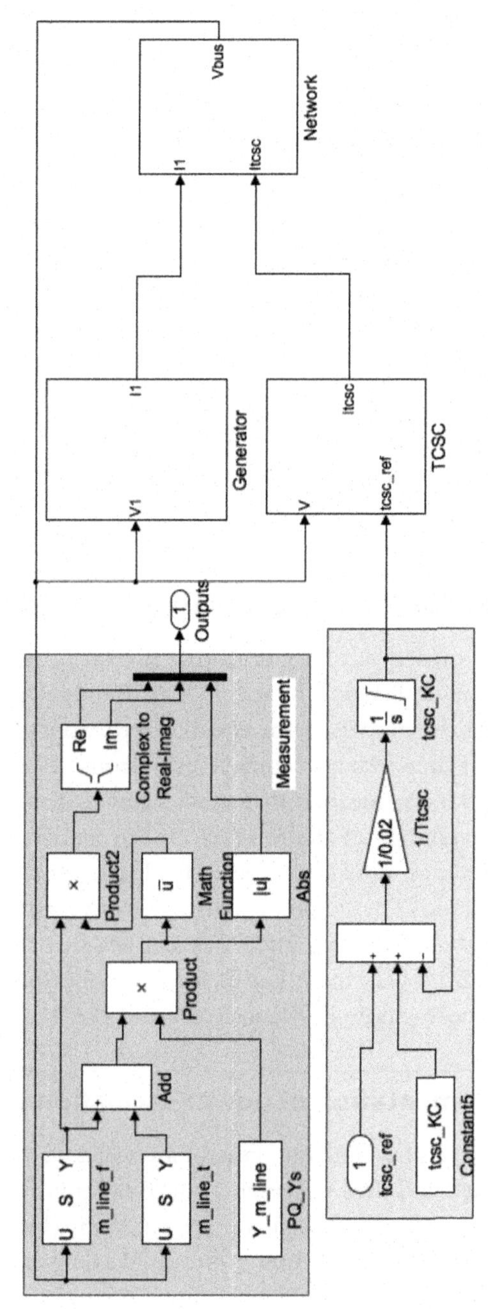

Figure 9.2 Simulink model of the two-area test system with thyristor controlled series compensator (TCSC).

This forms the list of feedback signals we are considering in our demonstration. Using the residue method, we will find the best signal to design a POD controller.

9.3.1.1 Feedback signal selection for power oscillation damping

As the feedback signals can be obtained from any part of the network, it is possible to come up with tens of signals or combinations of signals. But for our demonstration, we will pick active power, reactive power and magnitude of current through four transmission lines: lines connecting Bus 6 to Bus 7, Bus 7 to Bus 8, Bus 8 to Bus 9 and Bus 9 to Bus 10. From this list of 12 signals, we will select one feedback signal based on the magnitude of the residue.

Now build the Simulink block as shown in Fig. 9.2 and use Script 9.2 to build the program to initialize the model.

9.3.1.2 Linearization and calculation of residue

So far we have carried out linearization at several parts of the book. But here there is a small challenge. The IC block used to break the algebraic loop effectively breaks the model at time equal to zero. So if we try to linearize the model at that time instant, we may get a wrong result. Follow the procedure in Script 9.3 to linearize this model and to find the residue. Fig. 9.3 shows the magnitude of the residues for the transfer function between the TCSC reference input and selected feedback signals corresponding to the interarea mode. The first four signals are active power, followed by reactive power and current magnitude. We can see that the residue for signals 3, 4, 11 and 12 are very high. Let us select the fourth signal, which represents the active power flow between Bus 9 and Bus 10.

9.3.1.3 Implementation of power oscillation damping

Once the feedback signal and the actuator are selected, the plant transfer function is extracted and used to design a controller transfer function. A variety of methods for controller design are presented in Chapter 4, which are applicable for the work in this chapter. Now build a Simulink program as shown in Fig. 9.4. The part of Fig. 9.2 representing the measurement is converted to a subsystem as in Fig. 9.4. A selector is connected at the output of

```
% Follow Script 3.8 to get bus matrix, line matrix, mac_con matrix,
% and excitation parameters

% changes for TCSC
tcsc_k = 7;   % tcsc from bus
tcsc_m =8;  % tcsc to bus
tcsc_indx = 10; % index of the line in the Line matrix where tcsc is
% installed
tcsc_KC = 0.4;   % tcsc compensation ratio : xc/xl
tcsc_Ttcsc = 0.02;   % time constant for TCSC model

Z_TCSC = line(tcsc_indx,3)+1i*line(tcsc_indx,4);
XL_TCSC = 1i*imag(Z_TCSC);
RL_TCSC = real(Z_TCSC);
Nbus = size(bus,1);
ExpandTCSC = zeros(Nbus,1);
ExpandTCSC(tcsc_k) = -1;
ExpandTCSC(tcsc_m) = 1;

% modify series reactance of transmission line to reflect presence of
% TCSC
line(tcsc_indx,4) = line(tcsc_indx,4)*(1-tcsc_KC);


% Obtain load flow solution with TCSC
[Y] = form_Ymatrix(bus,line);
% calculate pre-fault power flow solution
[bus_sol, line_flow] = power_flow(Y,bus, line);

% find steady state voltage at the tcsc buses
tcsc_vk = bus_sol(tcsc_k,2)*exp(1i*bus_sol(tcsc_k,3)*pi/180);
tcsc_vm = bus_sol(tcsc_m,2)*exp(1i*bus_sol(tcsc_m,3)*pi/180);

% specify lines to make measurement
% [from bus to bus]

% Finding series impedance of transmission line
% for simplicity only consider the series admittance of the line
Y_m_line = -Y(m_line(:,1)+(m_line(:,2)-1)*size(Y,1));


% Synchronous machine initialisation

Smachs=[1;2;3;4];% buses with synchronous machines
Nbus=size(bus,1); % Number of buses
NSmachs=size(Smachs,1); % Number of synchronous machines


% program to initialise the synchronous machines
Synch_parameter_sys_base_fourmach
generic_Sync_Init

% Copy the code in the Script 3.9
```

Script 9.2 Program to initialize two-area system with thyristor controlled series compensator (TCSC).

```
% Open the Simulink model, Go to  ''Configuration Parameters > Data
% Import/Export > Format''  and change parameter set to 'Array',
% 'Structure' or
% 'Structure with time'. Close the dialogue box.


% type the following three lines of code into the MATLAB command
% window:
set_param(bdroot,'AnalyticLinearization','on')
set_param(bdroot,'BufferReuse','off')
set_param(bdroot,'RTWInlineParameters','on')

% In the Simulink model, click the Step Forward button once to
% compile the model, and then a second time to move it to a non-zero
% time.

% Obtain the state space representation using, For more information
% about these commands, type help linmod in the Matlab command window

X = Simulink.BlockDiagram.getInitialState('two_area_synch_tcsc_lin');
PARA = [1e-5, 0.001, 0];
[A,B,C,D] = linmod('two_area_synch_tcsc_lin',X,0,PARA);

% find eigenvalues and eigenvectors
[vright,d] = eig(A);
dr = diag(d);

% form left eigenvector
vleft = inv(vright);
vleft = transpose(vleft);

% Find index of the mode we are interested in
mode = find(real(dr)<1 & imag(dr)>3 & imag(dr)<5);

% Calculate residue.

R = C*vright(:,mode)*transpose(vleft(:,mode))*B;
```

Script 9.3 Program for linearization and residue calculation of two-area system with a thyristor controlled series compensator (TCSC).

the measurement block, which selects the fourth signal (active power flow from Bus 9 to Bus 10) of the 12 signal. The state space model represents the controller transfer function. The parameters of the state space model are `tcsc_cntr.a`, `tcsc_cntr.b`, `tcsc_cntr.c`, `tcsc_cntr.d`, and `tcsc_cntr_x` calculated using Script 9.4. The gain block is used to vary controller gain. The results presented in this chapter are obtained using a gain of 0.8. The POD can be deactivated by setting the gain to zero.

We need to make one more very important change in the Simulink block before we test the performance. Open the TCSC block. On the sum block connecting the tcsc_ref input, change the sign from positive to negative. This creates a negative feedback.

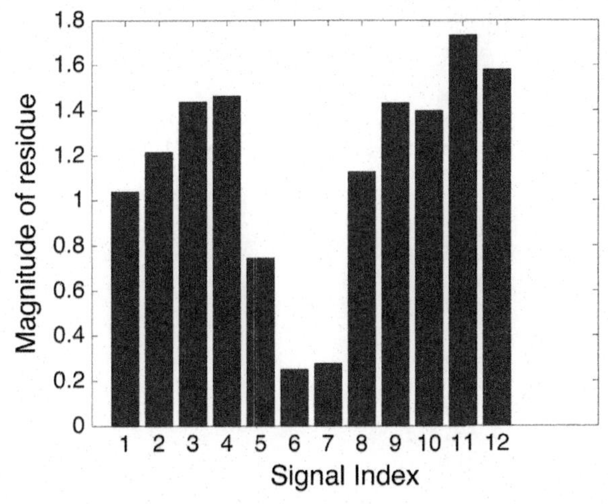

Figure 9.3 Residue magnitude for transfer function, the thyristor controlled series compensator (TCSC) reference input and selected feedback signals for the interarea mode.

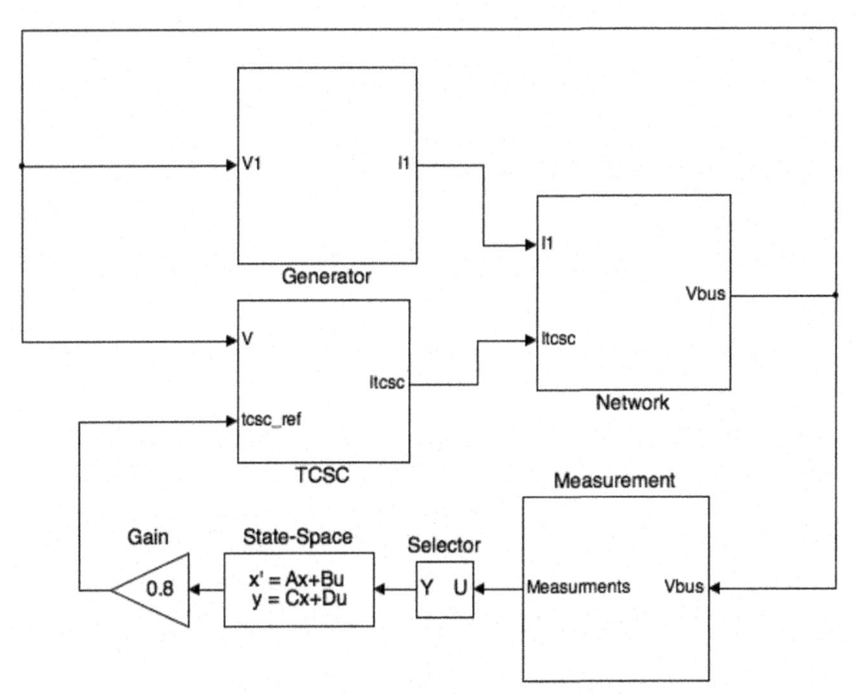

Figure 9.4 Simulink model of the two-area system with a thyristor controlled series compensator (TCSC) power oscillation damping (POD) controller.

```
% we have selected the active power flow between buses 9 and 10 as
% feedback signals. Lets find out its pre-fault active power flow
% through
% this line.
fbV = bus_sol(9,2).*exp(1i*bus_sol(9,3)*pi/180); % from bus voltage
tbV = bus_sol(10,2).*exp(1i*bus_sol(10,3)*pi/180); % to bus voltage
yft = Y_m_line(4);  % series admittance

% INITIAL ACTIVE POWER
Pft = fbV*conj((fbV-tbV)*yft);

% INPUT TO THE CONTROLLER AT TIME = 0,
u = real(Pft);

% Controller transfer function. Consisting of two transfer functions
tf1 = tf([10,0],[10 1]);
tf2 = tf([0.01 1],[0.6 1]);
tcsc_cntr = ss(tf1*tf2);

% converting controller transfer function to state space form
A = [tcsc_cntr.a; tcsc_cntr.c];
B = [tcsc_cntr.b; tcsc_cntr.d];

% finding initial value of the state space representation
tcsc_cntr_x=linsolve(A,-B*u);
```

Script 9.4 Thyristor controlled series compensator (TCSC) controller definition and initialization.

9.3.1.4 Controller performance

Controller performance can be compared by eigenvalues and using time domain simulation result cases with and without the POD controller. To simulate a case without POD controller, set the gain in the control loop to zero. For the case with POD controller, set a gain to 0.8. Later, the readers are encouraged to test the performance with other values of controller gains.

The eigenvalue plot showing the interarea mode of the two-area system with TCSC-POD is shown in Fig. 9.5. The controller improves damping of the interarea mode from 0.7% to 11%. The response of the system to a three-phase fault is shown in Fig. 9.6. The POD dampens the oscillations successfully. The bottom figure shows the TCSC compensation reference change following the fault. This value is capped between 0.05 (5%) and 0.5 (50%).

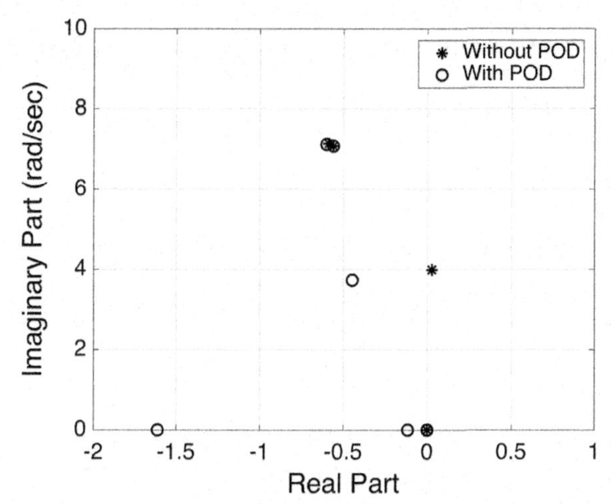

Figure 9.5 Eigenvalue plot showing the interarea mode for two-area system with a thyristor controlled series compensator (TCSC) power oscillation damping (POD).

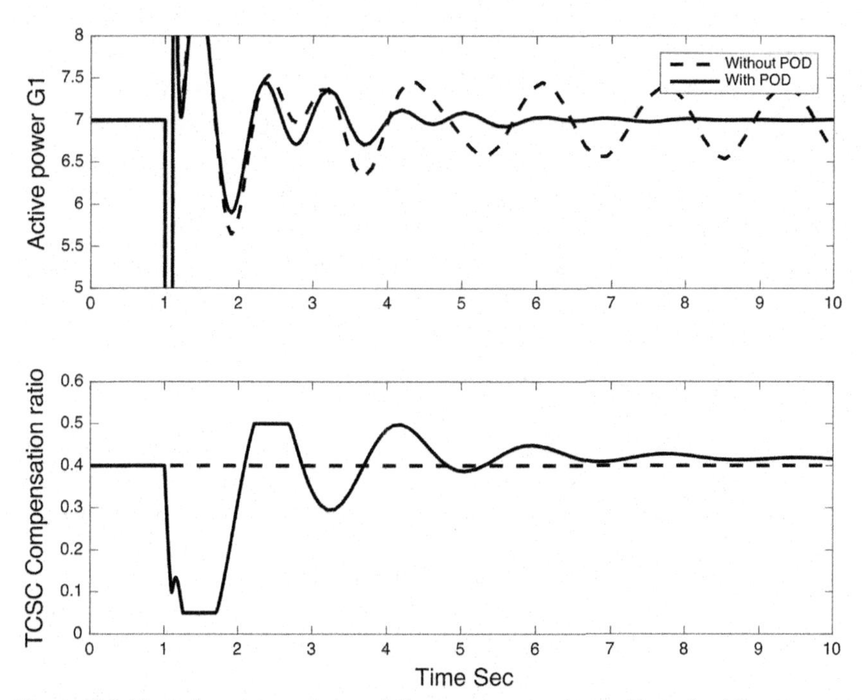

Figure 9.6 Time domain response of the two-area system with and without power oscillation damping (POD).

9.4 Two-area system with a static VAR compensator

In the last section, we tested a POD using a TCSC on a two-area test system model. In this section, an SVC is installed at Bus 9 and a POD controller is designed. To avoid repetition, discussion on linearization, residue calculation and feedback signal selection is reduced. This is because the theory and procedure are the same as we discussed for the TCSC.

In Chapter 8, we have developed an SVC model and tested it on an SMIB test system. The SVC block and initialization program are used to build the two-area system model here. Like in our previous cases, a three-phase fault is applied at time 1 s. All the 12 feedback signals considered for the POD design for the TCSC are used here as well.

Fig. 9.7 shows the Simulink model of the two-area system with an SVC. An input port is connected to the SVC supplementary input, and an output port is connected at the measurement block output. Follow Script 9.5 to initialize the model. Use the Script 9.6 to calculate the residue of the transfer function between SVC input and the measurements corresponding to the interarea mode. The magnitude of the residues for the 12 signals is shown in Fig. 9.8. The active power and magnitude of current measurements provide a good residue. For demonstration, let us select the active power flowing from Bus 9 to Bus 10 as the feedback signal.

Fig. 9.9 shows a Simulink model of the two-area test system with an SVC POD controller. The selector block parameter is set to select the fourth input out of 12. The state space block represents the controller transfer function, which is given in Script 9.6. The gain block controls the controller gain. If it is zero, the POD controller is inactive or system does not have a POD controller. When a POD controller is included, we use a gain of 0.06 in this demonstration.

The eigenvalue plot in Fig. 9.10 shows eigenvalues of the system with and without the controller. Fig. 9.11 shows the response of the system following a three-phase fault. When a POD controller is used, the oscillation settles within a few seconds demonstrating its effectiveness.

9.5 Two-area system with wind turbines

In this section, we will build a simulation model to analyze power system having synchronous machines and wind farms. Typically, a wind farm contains tens or hundreds of wind turbine generators, connected to a grid (the connection point is often called point of common coupling or PCC)

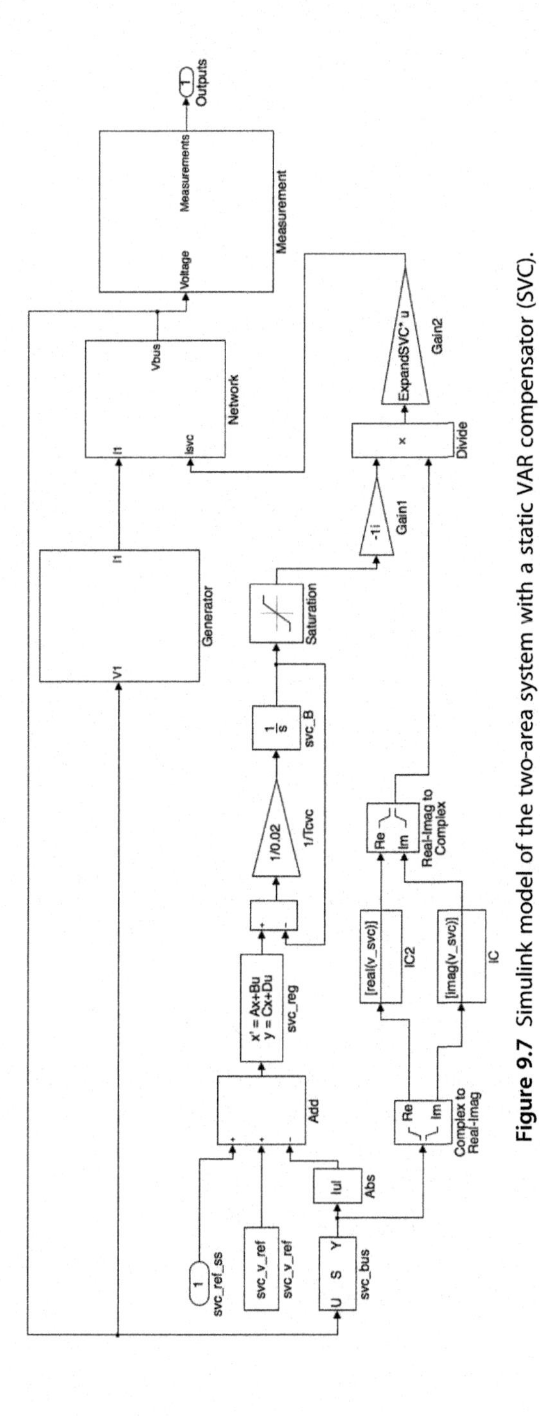

Figure 9.7 Simulink model of the two-area system with a static VAR compensator (SVC).

```
% Follow Script 3.8 to get bus matrix, line matrix, mac_con matrix,
% and excitation parameters

% changes for SVC
svc_bus = 9;
bus(svc_bus,5) = 0.5;  % specify reactive power output of SVC here
Nbus = size(bus,1);



% Obtain load flow solution with SVC
[Y] = form_Ymatrix(bus,line);
% calculate pre-fault power flow solution
[bus_sol, line_flow] = power_flow(Y,bus, line);

% Follow and Run script 8.4 for SVC initialisation

% specify lines to make measurement
% [from bus two bus]
m_line = [6 7; 7 8; 8 9; 9 10;];

% Finding series impedance of transmission line
% for simplicity only consider the series admittance of the line
Y_m_line = -Y(m_line(:,1)+(m_line(:,2)-1)*size(Y,1));

% Follow Script 9.3 to initialise synchronous machines

% Follow Script 9.2 to find post-fault impedance matrix
```

Script 9.5 Program to initialize the two-area test system with a static VAR compensator (SVC) at Bus 9.

through a medium voltage network called collector system. Modelling a complete wind farm is computationally intensive task, and for most power system studies, an aggregate model of a wind farm is sufficient. There are various approaches to making an aggregated wind farm. But those topics are not relevant at this instance. The objective in this chapter is to demonstrate how to put together a simulation model for a system with wind turbines. Hence, one of the synchronous machines in the two-area system is replaced with a wind turbine to make a study system for this section. The capacity and output of wind turbine is same as the synchronous machine. Once readers mastered the approach, a better representation of a wind farm can be used to make more realistic system studies. The wind turbine model used here can be either a DFIG type or PMSG type.

9.5.1 Building Simulink model

So far we have built all the building blocks such as synchronous machines, wind turbines and network. In addition, in the last sections, simulation of

```
% we have selected the active power flow between buses 9 and 10 as
% feedback signals. Lets find out its pre-fault active power flow
% through
% this line.
fbV = bus_sol(9,2).*exp(1i*bus_sol(9,3)*pi/180); % from bus voltage
tbV = bus_sol(10,2).*exp(1i*bus_sol(10,3)*pi/180); % to bus voltage
yft = Y_m_line(4);  % series admittance

% INITIAL ACTIVE POWER
Pft = fbV*conj((fbV-tbV)*yft);

% INPUT TO THE CONTROLLER AT TIME = 0,
u = real(Pft);

% Controller transfer function. Consisting of two transfer functions
tf1 = tf([10,0],[10 1]);
tf2 = tf([0.5 1],[0.65 1]);
svc_cntr = ss(tf1*tf2);

% converting controller transfer function to state space form
A = [svc_cntr.a; svc_cntr.c];
B = [svc_cntr.b; svc_cntr.d];

% finding initial value of the state space representation

 vc_cntr_x=linsolve(A,
```

Script 9.6 Static VAR compensator (SVC) controller definition and initialization.

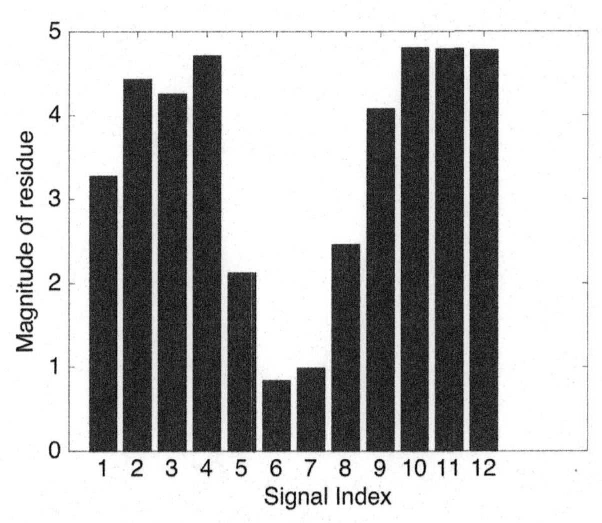

Figure 9.8 Magnitude of residue of the transfer function between static VAR compensator (SVC) input and the measurements corresponding to the interarea mode.

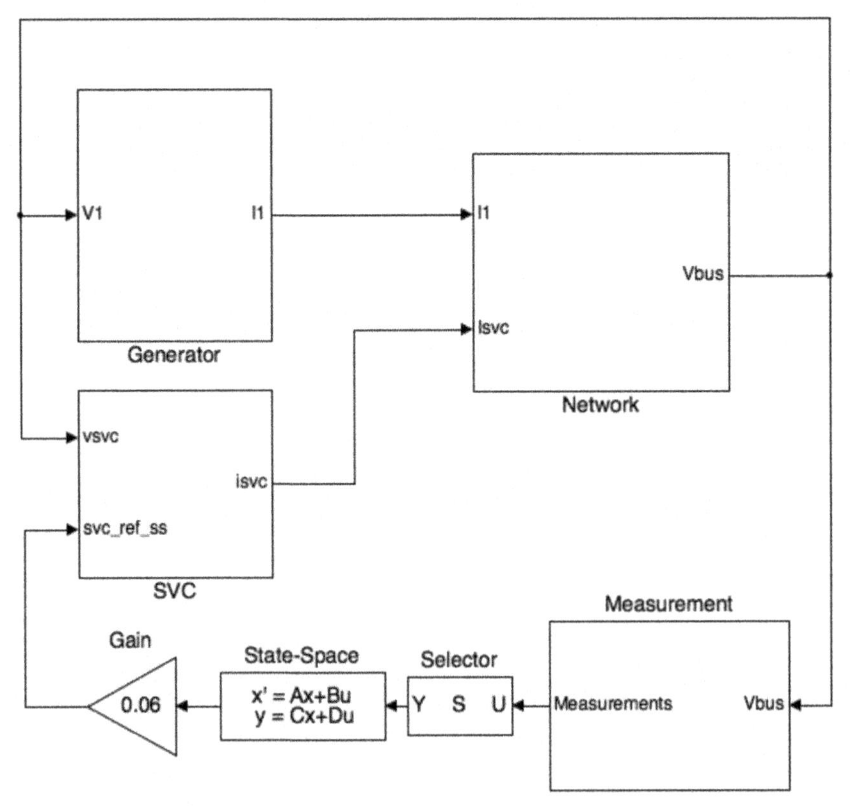

Figure 9.9 Simulink model of the two-area system with a static VAR compensator (SVC) power oscillation damping (POD) controller.

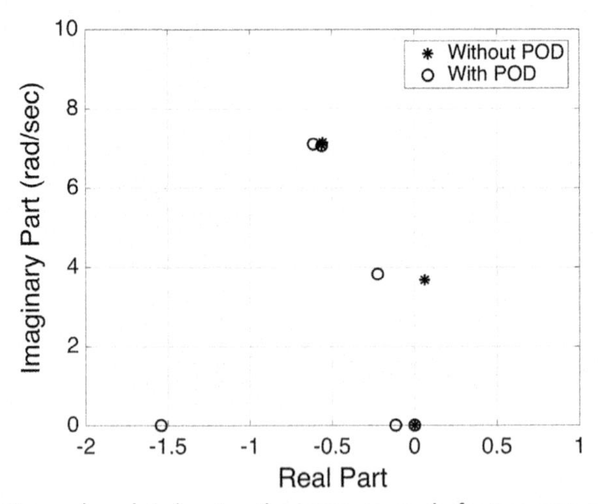

Figure 9.10 Eigenvalue plot showing the interarea mode for two-area system with a static VAR compensator (SVC) power oscillation damping (POD).

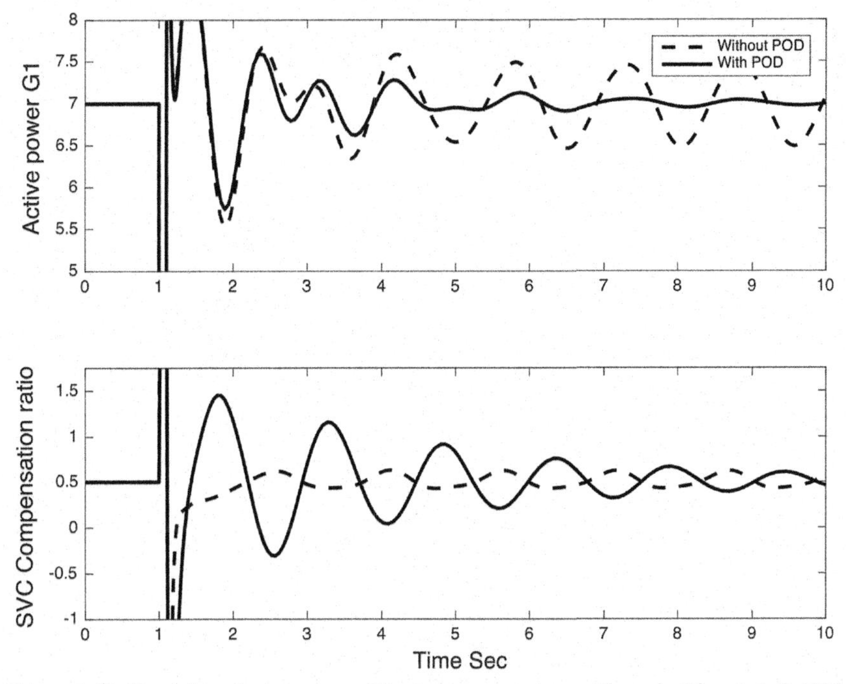

Figure 9.11 Time domain response of the two-area system with and without static VAR compensator (SVC) power oscillation damping (POD).

a system with synchronous machines and TCSC or SVC is demonstrated. Using the same approach, build a new Simulink model as shown in Fig. 9.12.

The model contains a synchronous machine subsystem, wind turbine subsystem and network subsystem. The wind turbine system can be either DFIG or PMSG subsystem developed in Chapter 6. Copy synchronous machine and network models from the previous sections. There are three gain blocks in the program, which needs some explanation.

Gain 3 with value 900/100: Rating of wind turbine is 900 MVA, same as that of the synchronous machines in the two-area test system. Intialization of DFIG or PMSG is carried out in machine base. Simulation program developed so far assumes system base MVA as 100 MVA. So output current of wind turbine block should be multiplied by a factor machine base/system base, which is equal to 900/100 in this case before connecting to network block.

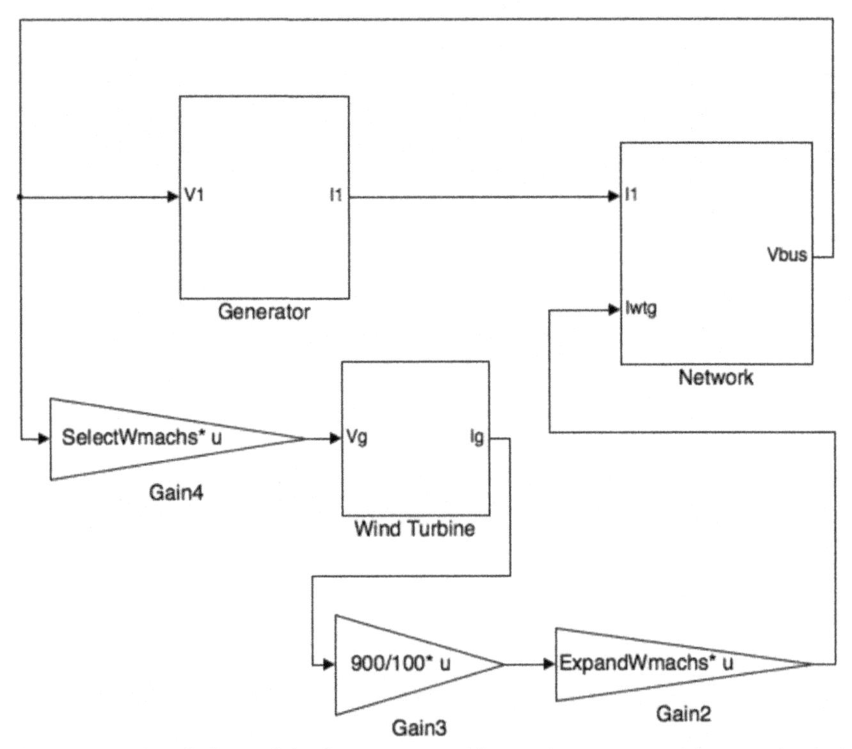

Figure 9.12 Simulink model of a system with synchronous machines and wind turbines.

SelectWmachs: This has similar meaning as SelectSmachs for synchronous machine, which selects appropriate voltage for the wind turbine block.

ExpandWmachs: This has similar meaning as ExpandSmachs, which expands wind turbine block output current to a vector of size equal to the number of buses in the network.

9.5.2 Initialization program

Now let us assemble initialization program for the two-area system with three synchronous machines and a wind turbine. Follow instructions in Script 9.7. The functions used in the script are developed in the previous chapters. As discussed before, the synchronous machine at Bus 4 is replaced with wind turbine. Depending on which type of wind turbine is required, make Pmachs = 4 indicating PMSG type wind turbine or Dmachs = 4 for DFIG-type wind turbines. (When the program is up and running, why not try to develop a program with two synchronous machines, a DFIG-type wind turbine and a PMSG-type wind turbine).

```
% Follow Script 3.8 to get bus matrix, line matrix, mac_con matrix,
% and excitation parameters

% Obtain load flow solution
[Y] = form_Ymatrix(bus,line);
% calculate pre-fault power flow solution
[bus_sol, line_flow] = power_flow(Y,bus, line);


% remove one synchronous machine
mac_con(4,:) = [];

Smachs=[1;2;3];% buses with synchronous machines
Nbus=size(bus,1); % Number of buses
NSmachs=size(Smachs,1); % Number of synchronous machines

% program to initialise the synchronous machines
Synch_parameter_sys_base_fourmach
generic_Sync_Init

% A new variable bus_sln is used becuas some of the program written
in
% Chapter 6 uses bus_sln instead of bus_sol
bus_sln = bus_sol;

% This is an important change. The power flow is obtained by assuming
% a system base value of 100 MVA. But rating of machine at bus 4 is
% 900 (same as the synchronous machine). So we must convert active
% and reactive power output of wind turbine to pu value at 900MVA
% base. This conversion is performed in the next line.

bus_sln(4,4:5) = bus_sln(4,4:5)*100/900;

% Specify Pmachs = 4 for PMSG turbine or Dmachs =4 for DFIG
Pmachs = []';      % Buses where PMSG is connected
Dmachs = [4]';      % Buses where DFIG is connected
Omega = 2*pi*50;

% initialization code for PMSG turbine
if size(Pmachs,1)

    find_pmsg_state_initial_conditions
    pmsg_mult = zeros(size(bus_sol,1),size(Pmachs,1));
    pmsg_mult(Pmachs,:) = eye(size(Pmachs,1));
    Wmachs = Pmachs;
end

 % initialization code for DFIG turbine

if size(Dmachs,1)

    find_dfig_state_initial_conditions
    dfig_mult = zeros(size(bus_sol),size(Dmachs,1));
    dfig_mult(Dmachs,:) = eye(size(Dmachs,1));
    Wmachs = Dmachs;
end

NWmachs = size(Wmachs,1);

% made ExpandWmachs and SelectWmachs matrices

ExpandWmachs= zeros(Nbus, NWmachs);
for counter=1:NWmachs
ExpandWmachs(Wmachs(counter),counter)=1;
end

SelectWmachs= zeros(NWmachs,Nbus);
for counter=1:NWmachs
SelectWmachs(counter,Wmachs(counter))=1;
end

% Copy the code in the Script 2.5 to represent Network

% If Zf is used in the network model, make fault impedance equal to
% pre-fault impedance to remove a network fault.
Zf = Z
```

Script 9.7 Program to initialize a two-area system with three synchronous machine and a wind turbine.

9.5.3 Simulation results

Figs. 9.13 and 9.14 shows active power output of the synchronous machines and wind turbine following a step change in wind speed in the two-area system with DFIG-type wind turbine and PMSG-type wind turbine,

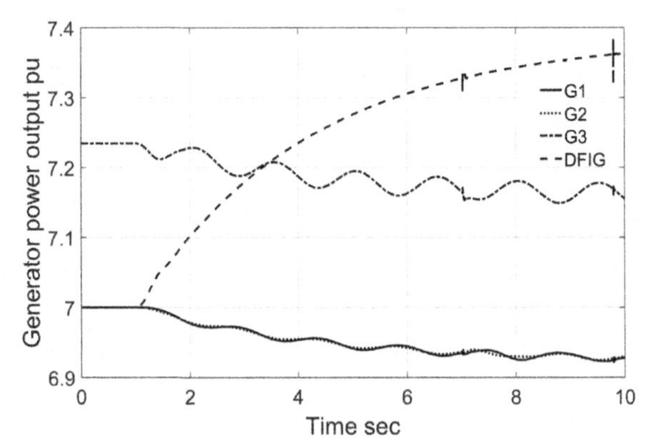

Figure 9.13 Time domain response of the two-area system with DFIG wind turbine following a 0.25 m/s increase in wind speed at time = 1 s.

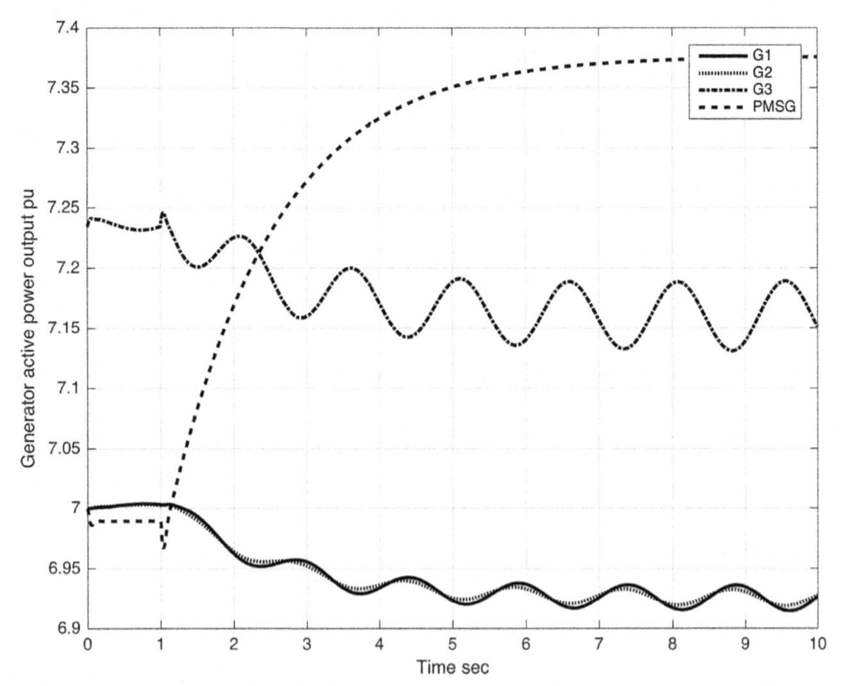

Figure 9.14 Time domain response of the two-area system PMSG wind turbine following a 0.25 m/s increase in wind speed at time = 1 s.

respectively. Use these results to validate your model. Depending on the type of wind turbine, the response of the system varies. However, do not try to make any conclusion from these results, as the aggregate wind farm representation used in this simulation is not accurate. Our objective is to demonstrate how to quickly build a Simulink model using programs already developed in the previous chapters. Readers should carry out further extension of the model using more realistic assumption on wind farm representation for any kinds of stability analysis.

9.6 Conclusions

In this chapter, a basic step to build a simulation program for system with synchronous machines, SVC, TCSC and wind turbine is described. It is shown that programs and building blocks developed in previous chapters can be put together to quickly develop a simulation program for complex systems. It should be noted that the objective of this chapter, in fact other chapters as well, is to demonstrate how to assemble various equations to build power system simulation programs. In doing so, authors have used many assumption, which are sometimes not realistic. For example, representing a 900 MW wind farm as a generator and a transformer helped us to demonstrate how to build the program quickly, but it is not a realistic way of modelling a large wind farm. Once readers have learnt how to build these programs, they are advised to refer recent literatures on power system modelling to find more accurate representation.

References

Kundur, P., 1994. Power System Stability and Control. McGraw-Hill.
Rogers, G., 2000. Power System Oscillations. Kluwer Academic Publishers.

Index

Note: 'Page numbers followed by "f" indicate figures, "t" indicates tables'.

T

U

V

W